W9-CNW-372

DRINK
YOUR WAY TO
GUT
HEALTH

140 DELICIOUS
PROBIOTIC
SMOOTHIES
& OTHER DRINKS THAT CLEANSE & HEAL

MOLLY MORGAN, RD, CDN, CSSD

Houghton Mifflin Harcourt
Boston New York 2015

Library of Congress Cataloging-in-Publication Data
Morgan, Molly.
  Drink your way to gut health : over 140 delicious probiotic smoothies and other drinks that cleanse and heal / Molly Morgan, RD, CDN, CSSD.
    Pages cm
  Includes index.
  ISBN 978-0-544-45174-2 (trade paper); 978-0-544-45176-6 (ebk)
Probiotics. 2. Smoothies (Beverages) —Therapeutic use. 3. Beverages— Therapeutic use. 4. Gastrointestinal system —Microbiology. I. Title
  RM666.P835M67 2015
  641.2—dc23

                        2014025821

A Hollan Publishing, Inc. Concept
Copyright © 2015 by Molly Morgan
Photography © 2015 by Linda Xiao
Food styling by Molly Shuster
Prop styling by Maeve Sheridan
All rights reserved.

For information about permission to reproduce selections from this book, write to Permissions, Houghton Mifflin Harcourt Publishing Company, 215 Park Avenue South, New York, New York 10003.

www.hmhco.com

Printed in the United States of America
DOC 10 9 8 7 6 5 4 3 2 1

To my taste testers
and supporters—
my husband and our
two little guys.

TABLE **OF** CONTENTS

acknowledgments

I am forever thankful to my patient, loyal, and caring husband for his ongoing support. Especially for his encouragement throughout the development of this book (from the days when our refrigerator looked like a science experiment), for his recipe ideas, and for discussing the volumes of research to support this book. Next, thank you to our two sons, who at times may have grown tired of listening to the sound of the blender, yet happily tasted and provided valuable feedback and ideas that were incorporated into the recipes. Lastly, I need to thank my wonderful parents, in-laws, and family for their always enthusiastic backing.

This book went from concept to reality thanks to Holly Schmidt of Hollan Publishing; without her the book would not have been possible. Thank you to Justin Schwartz and Houghton Mifflin Harcourt for their trust in me and commitment to publishing this book. Moreover, to all of those who had a part in making the content of this book come together, from the editors and photographers, to the designers —thank you!

And thank you to the readers who try these recipes as they drink their way to gut health.

introduction

The gastrointestinal (GI) tract is a complex ecosystem and home to trillions of microbes (aka bacteria)—some good and some bad. There is emerging research to show that gaining a better understanding of the microbes—good and bad—that reside in the GI tract could unlock deeper understanding and new ways to diagnose and treat a wide variety of diseases.

Our ancestors for tens of thousands of years were unknowingly consuming fermented foods and beverages. Over time, they began to recognize the palatability, preservative nature, and other properties that these foods delivered. While probiotics have a long history of health-helping connections, the term only appeared recently: It was first used by Daniel Lilly and Rosalie Stillwell in 1965 to describe substances secreted by one microorganism that simulate the growth of another.

Today probiotics are often referred to as "good bacteria," and are increasingly being linked to health benefits, including improving gastrointestinal health, helping conditions like psoriasis and chronic fatigue syndrome, and even potentially playing a role in obesity and mental health. The microbes in our intestines also provide us with a barrier to infection, provide metabolic fuel, and contribute to normal immune development.

There are basic probiotic-rich ingredients that form the foundation of the recipes in this book, including buttermilk, yogurt, cultured nondairy yogurt, filmjölk, frozen yogurt, kefir, kombucha tea, ginger beer, miso, and tempeh. It has been found that there are also small amounts of probiotics in soy milk and almond milk. In this book you will find DIY recipes for many of these ingredients, including almond milk for making nondairy versions of yogurt, as they will serve as the base for other beverages.

There are other foods that are probiotic rich, including natto, sauerkraut, sourdough bread, pickles, soft cheese (like Gouda), kimchi, and olives. Although you will find only one of the foods (the pickles in "Jose Gerbousky," page 228) in the drink recipes, try working them into your eating routine for an intestinal boost!

food first, supplements second

In response to the growing interest in probiotics, there is an increase in the number of probiotic supplements popping up on the market. When it comes to taking supplements, remember that a dietary supplement is defined as a product intended for ingestion that contains a "dietary ingredient" intended to add further nutritional value to (supplement) the diet. The key piece of the definition is *to add further nutritional value*—in other words, you must first consume the nutrients your body needs from foods, and supplements are intended to add further nutritional value, not to become a crutch for poor eating habits.

First get probiotics from your food and beverages, then if you still need to add further probiotics, work with a registered dietitian and your health care provider to carefully consider adding a supplement. What you will quickly realize is that there is a wide variety of ways to add probiotics to your daily routine—if you don't like yogurt, try kefir; if you don't prefer kefir, try ginger beer. And always remember, it can take time to learn to like new foods, flavors, and textures—by being adventurous, I am confident that anyone can find probiotic foods and beverages to add to their eating routine.

the prebiotic connection

There is another group of foods, called prebiotics, that are linked to improving overall intestinal health. They are non-digestible and non-absorbable carbohydrates that help feed the good bacteria that are already living in the digestive system.

Examples of prebiotic foods include: bananas, asparagus, leeks, onion, garlic, almonds, pistachios, red wine, honey, maple syrup, oatmeal, whole grains, and legumes. Throughout the recipes in this book, you will find many prebiotic foods added to the recipes to help boost the potential power of the probiotics in your digestive tract.

Research has found benefits from a wide range of prebiotics, including: fructo-oligosaccharides (FOS), polyfructan inulin, galacto-oligosaccharides (GOS), and lactulose. For example, FOS has been found to increase calcium absorption; food sources include wheat, barley, bananas, asparagus, tomatoes, onions, leeks, garlic, and agave. And lactulose is widely used to treat constipation.

Other potential prebiotics include lactitol, xylo-oligosaccharides, isomalto-oligosaccharides, soybean oligosaccharides (raffinose and stachyose), lactosucrose, resistant starch, and cereal fibers.

the health connection

The connection between probiotics and health seems to lie within our intestinal tract, which is inhabited by trillions of microbes.

Early discovery of probiotic benefits is credited to the Russian scientist Elie Metchnikoff for his work on the positive role that probiotics have on health. In his book *The Prolongation of Life* (1907), he suggested that foods like yogurt, kefir, and sour milk (containing lactic acid bacteria) were associated with good health and longevity. His reports were based on the Bulgarian peasants who consumed large quantities of sour milk, and lived longer than the average population.

Studies show that a "Western" diet that is high in fat and sugar results in a more porous intestinal lining, which results in systemic access to food antigens, environmental toxins, and more. Adding probiotics can positively boost intestinal health by improving the bioavailability of vitamins and minerals like B-vitamins, magnesium, and zinc.

To date, the United States Food and Drug Administration (FDA) has yet to approve a health claim for probiotics, although a few other health agencies around the globe have, including Japan. Their National Institute of Health, Labour, and Welfare, under the Foods for Specified Health Use (FOSHU) program, has identified 65 probiotic products containing one or more of the 16 different approved probiotic strains including several strains of *Lactobacillus* and *Bifidobacterium*. The health targets for probiotics in Japan include gastrointestinal conditions, immunity, allergy, cold and flu-like symptoms, cholesterol levels, blood pressure levels, and diabetes.

All probiotics do not produce the same health benefits. For example, *Lactobacillus rhamnosus* has been found to be effective in the treatment of rotavirus diarrhea in children, while another strain had no clinical impact.

Below is a brief overview of groups of probiotics and/or particular strains that have been linked by research to improve health:

- Improved digestion: *Bifidobacterium* and *Lactobacillus*
- Immune stimulation: *Bifidobacterium* and *Lactobacillus*
- Increased bioavailability of iron: *Lactobacillus acidophilus* SBT2062
- Reduced cholesterol: *Lactobacillus*
- Reduced respiratory infections: general intake of probiotics
- Decreased dental carries in children: *Lactobacillus rhamnosus* GG
- Reduction in acute (rotavirus) diarrhea in children: *Lactobacillus rhamnosus, S. boulardii*
- Possible influence on brain health: *Lactobacillus* and *Bifidobacterium*
- Sustainable weight loss in women: *Lactobacillus rhamnosus*

Additionally, research is linking gut microbe diversity to reduced obesity, improved blood glucose control, reduced insulin resistance, and mental health benefits:

- Blood sugar control: Studies show that the addition of probiotics can improve fasting blood sugar levels.
- Mental health: There is a growing area of mental health, what is becoming known as *nutritional psychiatry,* that is finding connections between foods and mental health. As it relates to probiotics, traditional Japanese dietary practices, which include a lot of fermented soy products, have been linked to lower rates of depressive symptoms.
- Obesity: There seems to be an association with a shift in the microbe diversity of the gut and obesity. Research suggests that the variety and types of the bacteria in the gut may have an impact on fat storage and in turn, the development of obesity. Studies have shown that a restriction of calories and a loss of weight in obese individuals resulted in a change in the gut microbes, for the better.

While scientists are still determining which strains of microbes have the biggest impact on health status, what we know is that good bacteria can ultimately help our health. Enjoy the recipes in this book as a way to improve your intake of probiotics and potentially impact your gut health.

equipping your kitchen

You don't have to spend a fortune to start making your own probiotic-rich beverages at home. Many of the supplies you will need are likely items you already have around; here is a list of the basics to get you started.

tools and gadgets

Pots and pans: You will need a small (4-quart) soup pot to boil milk and water to prepare kefir, yogurt, and kombucha.

Thermometer: Working with active cultures and kefir grains requires you to know the temperature of the liquids.

Blender: You'll need a standard kitchen blender or high-powered blender to blend the smoothies. Note: Some recipes work best with a high-powered blender (like a Ninja or Vitamix), although they can still be made with a standard kitchen blender or a food processor.

Quart jars or containers: Keep a number on hand for storing and fermenting beverages.

Cheesecloth or coffee filters: Either is good for covering the jars while fermenting beverages.

Rubber bands: These are needed for securing cheesecloth or coffee filters to the containers during the fermenting process.

Yogurt incubator (aka yogurt maker): A yogurt maker will take yogurt prep time down to minutes. The yogurt recipes in this book include steps to make yogurt without a yogurt maker too.

Yogurt strainer: You can make your own Greek yogurt with a yogurt strainer or you can use cheesecloth.

For more equipment recommendations, see Resources, page 247.

the base ingredients

There are several ingredients used in the recipes that provide health-helping probiotics. Most of these products are widely available in grocery stores, and throughout the book there are recipes to make many of them yourself. For some ingredients you may need to go to a specialty health food store or the "natural" or health food section of the grocery store.

Buttermilk is milk that has been cultured with lactic acid to form cultures and probiotics. It is low in calories with just 100 per cup. Although butter is in the name, it actually does not contain any butter.

Cultured almond milk yogurt is cultured from milk made from ground almonds mixed with water. Almond milk is naturally cholesterol free and low in protein compared to cow's milk.

Cultured coconut milk is cultured from coconut "milk" (made from the white pulp of coconuts that is gently pressed to release the flavorful liquid). Also called cultured coconut milk yogurt, it is cultured through the same process as regular yogurt.

Filmjölk is a traditional drinkable yogurt from Sweden that is slowly fermented using *Lactococcus lactis* and other live cultures to provide a soft butter-like flavor. Filmjölk can be used as a substitute for kefir, yogurt, or buttermilk in recipes.

Greek yogurt is a thicker style yogurt that has been strained to remove the whey (clear liquid), which in turn results in a thick yogurt with a distinctive, sour taste and higher protein content.

Kefir comes from the Turkish word *keyif,* which means "good feeling." It is a thick, tangy, yogurt-like beverage that contains a wide variety of probiotic bacteria. Kefir is made from the addition of kefir grains to milk; bacteria, yeasts, and proteins in kefir grains work together to produce kefir. The three main types of probiotics in kefir are *Lactobacilli, Lactococci,* and *Leuconostoc.*

Kefir water is a fermented beverage made from water, sugar, and water kefir grains. The bacteria in the water kefir grains then metabolize the sugar and produce beneficial bacteria.

Kombucha is a fermented tea drink that is traditionally made from sweetened black tea that is fermented by a SCOBY (symbiotic colony of bacteria and yeast), aka kombucha mushroom. Different teas are often used to provide different flavor profiles.

Lassi is a probiotic beverage popular in India and Pakistan that is traditionally savory in nature and consists of a blend of yogurt, water, spices, and sometimes fruit.

Miso is a fermented soybean paste that is believed to have originated in China in 600 AD or earlier. It is a thick paste-like substance produced by fermenting soybeans with salt and the fungus *kojikin.* It is a traditional salty and tangy Japanese seasoning used as a base for soup, sauces, and vegetable dishes.

Pickles are cucumbers (or other vegetables or fruits) that are preserved in brine or vinegar. All pickles are not created equal when it comes to probiotics. The types of pickles that are rich in probiotics have traditionally been pickled without vinegar, so look for varieties that are processed without it.

Soy milk is made from soy beans and naturally contains some probiotic benefits.

Soy milk yogurt is made from cultured soy milk. It has been found that cultured soy milk has enhanced the availability of the soy isoflavones. Additionally, the fermentation process assists in protein digestion, enhances intestinal health, and supports a healthy immune system.

Tempeh is an Indonesian food that is made from fermented soy beans or other grains and is a source of protein with a nutty flavor.

Yogurt is a fermented milk product that is made from cow's milk.

the ingredient basics

frozen fruit

Frozen fruit is in many of the beverages and smoothie recipes. From week to week, vary the types of frozen fruit you buy to switch up the variety of flavors and nutrients.

Bananas whip up nicely in smoothies and provide a slightly sweet creamy texture to drinks. Freezing bananas is perfect for those times when you may have bought too many bananas! To freeze them, simply remove the peel, break the banana into smaller pieces, and place in a labeled freezer bag.

Tip: When purchasing frozen fruit, look for frozen whole or sliced fruit varieties and skip those with added sugar or sauces.

frozen vegetables

Some frozen vegetables, like carrots and butternut squash, work well in smoothies.

Tip: When purchasing frozen vegetables, opt for varieties without added sauces.

milk

Many of the recipes in the book include milk—some specify a particular type of milk, others just list "low-fat or nonfat milk." For those, choose what type of milk you prefer—almond milk, light almond milk, soy milk, rice milk, coconut milk, cow's milk, etc.

Almond milk vs. regular milk: Some of the recipes in the book specify light almond milk because it has only 40 calories per cup. The downside is that almond milk is lower in protein, with only 1 gram per cup compared to its cow's milk counterpart (8 grams per cup). Yet, most people tend to get plenty of protein in their daily food intake so the swap can help lighten up the calories while still adding a slight creamy flavor. But keep in mind that regular milk (cow's milk) can be used in recipes that call for almond milk.

Coconut milk vs. coconut milk beverage: Coconut milk is literally the "milk" from pressed coconut meat blended with water (a 1:1 ratio). Coconut milk beverage is coconut milk that is blended with water, sweetener, and thickeners. In the recipes that include coconut milk, either could be used, although nutritionally, coconut milk beverage (even though it is sweetened) is lower in calories and fat compared to coconut milk.

Note: Coconut milk is typically sold in cans and coconut milk beverage is sold in refrigerated section of the grocery store or in shelf-stable boxes or cartons. Many of the recipes in the book call for light coconut milk simply because it is lower in calories than regular coconut milk; light coconut milk has only 45 calories per ¼ cup compared to the 100 calories of regular coconut milk.

yogurt

Many recipes here call for plain yogurt so you can control the sweetness of the drink rather than starting with the sweetness that comes with flavored yogurt. Making your own yogurt is super simple, especially with a yogurt incubator (aka maker), and it can actually save you a bundle of money, once you recoup the initial purchase of the yogurt maker. For more on making your own yogurt, see the recipe on page 44.

Note: If you have a flavored yogurt on hand that matches well with the flavors in the drink or smoothie, it could be used as a substitute for the plain yogurt; just skip adding any additional sweeteners as flavored yogurt already has added sugar.

Greek vs. regular yogurt: From a nutrition perspective, Greek yogurt has significantly more protein than regular yogurt and a thicker texture. If you prefer a protein boost in the recipes, swap in Greek yogurt for those that call for regular yogurt. But note that you may need to increase the amount of liquid in the recipe on account of the thicker texture of Greek yogurt. Any regular yogurt can be turned into Greek yogurt by straining it; see the Resources on page 247 for more details.

Nondairy options: From coconut milk to almond milk there are many cultured nondairy yogurt options available commercially. Making your own nondairy yogurt is also an option, although it takes a little more practice than making your own cow's milk yogurt because it is more watery in texture.

kefir

Making your own kefir requires just a few supplies. A bonus is that you can choose what milk you want to use to make kefir and get started. If you opt not to make your own kefir, bottled kefir is available at grocery stores in a range of flavors from plain to strawberry or peach. Just like yogurt, most recipes use plain kefir as the base, then you can adjust the sweetness to your preference.

kefir water and coconut kefir water

Kefir water is made with specific water kefir grains. There are commercially available flavored kefir water beverages and coconut kefir water, which can be used in the recipes in this book in place of the Water Kefir recipe. See Resources, page 247, for more details.

kombucha

Made through a process of fermentation, kombucha is available in many brands and options. In addition, there are several online options for purchasing kombucha brewing supplies and ready-made kombucha. See Resources, page 247, for more details on kombucha supplies. If you choose to buy kombucha rather than brew your own, notice the flavor variety that you are buying to make sure it will match well with your recipe.

ginger beer

The simple brewing process used to make ginger beer includes ginger, yeast, sugar, and water. Making ginger beer takes only a matter of minutes to put together, then 24 to 36 hours to ferment and get bubbly. Depending on your preference, you can adjust the quantity of ginger used for a milder or stronger ginger flavor. Store-bought varieties are available in grocery stores.

Ginger beer vs. ginger ale: Ginger beer is a fermented, brewed beverage that is naturally carbonated through the brewing process. This is much different from ginger ale, which is just carbonated water that is flavored with ginger and sugar; ginger ale does not contain probiotics.

sugar

Some of the recipes call for a specific type of sugar (agave nectar, honey, maple syrup) because the flavor profile of the sweetener works well with the recipe. Some recipes will just list "sugar" as the ingredient and that can be a sweetener of your preference such as granulated sugar, raw sugar, agave nectar, or honey. Even more importantly than what type of sugar you choose is to keep the amount of sugar you add to a minimum.

While some sugar varieties may have health-helping properties or have a lower glycemic index, they are still sugars and need to be used in moderation. Some of the recipes in this book do not add sugar, others add a modest amount, and some need the sugar as "food" for the fermentation process (like ginger beer and kombucha). When you are mixing up a smoothie recipe that calls for sweetening, try tasting the drink first without adding any extra sweetener and then gradually add sugar to your taste preference, or skip it all together.

Sugar substitutes: The recipes in this book do not use sugar substitutes like stevia, aspartame, or sucralose; I do this because although they would lower the sugar content of a recipe, they still come with a super sweet taste. Overall, in my opinion, getting hooked on super sweet taste—whether it is from "real" sugar or sugar substitutes—is something to avoid.

safety and tips

hand washing 101

The first step to preparing safe and healthy food is proper hand washing. While it seems like a simple step, if overlooked you can spread harmful bacteria to the foods you are preparing. *Before* you begin making these recipes that are rich with healthy bacteria, follow these steps for proper hand washing to keep the harmful bacteria out.

- Rinse your hands under warm water then apply soap.
- Rub your hands together vigorously and scrub between your fingers and under the nails! Don't forget the back of your hands, as bacteria can hang out there too.
- Then continue scrubbing your hands for at least 20 seconds.
- Rinse your hands under running water and then dry well with a clean dry towel.

kitchen cleaning

Make certain to keep all surfaces and cooking utensils clean because harmful bacteria can easily spread around the kitchen.

- Prepare countertops by washing them with hot soapy water before you get started.
- Wash your dishcloths and kitchen towels often on the hot cycle of the washing machine.

preparing fruits and vegetables

Always wash fruits and vegetables first! *Even* if you are planning to peel a fruit or vegetable, scrub and wash the outside of it first. Then cut away any damaged or bruised areas on the produce.

To wash produce: Wash under running water. For thick-skinned produce like oranges, cucumbers, and melons, use a clean produce brush. Pat dry with a paper towel or clean cloth.

What about bags of spinach marked "pre-washed"? Although it is safe to use without further washing, to be on the safe side, it is always best to wash it before use.

storing

After the fermented beverages and products have completed the fermentation process, store them in a refrigerator. It is best to consume the smoothie immediately for optimal freshness, or store in the refrigerator and consume within 3 or 4 days.

don't be afraid to toss a bad batch

When you are beginning to brew these beverages and make these products at your home, you may have some batches that flop. You will know this almost instantly! For example, my first batch of homemade yogurt kefir, after fermenting for 24 hours, had an awful smell… something did not go right with the process. What to do? Discard the batch and start again. Although it may seem wasteful, it is necessary because consuming improperly brewed probiotic beverages could be harmful.

nutrition facts

Each recipe in this book has the Nutrition Facts information listed per serving, which includes a percentage for vitamin A, vitamin C, calcium, and iron. This number is for the percentage of the "daily value" (DV) of a nutrient—the amount of the nutrient (set by the FDA) the average person should consume each day for optimal health. Foods providing 20 percent or more of the DV are considered to be high sources of a nutrient, but foods providing lower percentages of the DV also contribute to a healthful diet.

Daily Values

Calcium: 1,000 mg for adults and children aged 4 years and older
Vitamin C: 60 mg for adults and children aged 4 and older
Vitamin A: 5,000 IU for adults and children age 4 and older
Iron: 18 mg for adults and children age 4 and older

BASIC
RECIPES

Making your own probiotic-rich beverages takes a little practice, but once you get the basics down, it goes very smoothly. The best part is that you can save a bunch of money and have a stock of probiotic-rich beverages on hand to drink your way to gut health. I suggest you tackle these basic recipes one at a time and remember that you can purchase store-bought versions if making your own seems too intimidating. Although, once you get started, my prediction is that you will be hooked, as it is fun and rewarding!

about kombucha

Kombucha is a fermented tea beverage that is said to have origi-nated in China around 220 BC, later spread to Russia before 1910, and then on to Europe. Some call kombucha the elixir of health and a cure-all for everything from arthritis to chronic fatigue syndrome. Although there is limited scientific evidence to support the claims that are touted about kombucha, it is a viable way to drink your way to gut health and introduce probiotics to your body.

How is kombucha made? It is made from fermenting tea (typically black tea) with a symbiotic colony of bacteria and yeast (SCOBY), aka kombucha mushrooms.

A word of caution: When preparing kombucha at home, it is im-portant to follow all kombucha brewing recommendations and steps. If there is any suspicion that the batch of tea you are making isn't good—for example, the appearance of little white or black spots over the top of the SCOBY (a sign of mold)—then you must discard the batch and the SCOBY and start over.

While kombucha is not routinely reviewed by the Food and Drug Administration (FDA), following the unexplained illness of two women (one resulting in death) in 1995, the FDA was prompted to review the situation. They evaluated the practices of commercial kombucha SCOBY production (aka kombucha mushrooms), and found no issues of pathogenic organisms or hygiene violations. It's important to note that one of the women who fell ill had increased her daily kombucha intake from 4 ounces to 12 ounces, and that over a hundred others were consuming kombucha made from the same SCOBY and did not fall ill.

This brings up two important points for those starting to brew their own kombucha and/or consume kombucha:

Buy your starter SCOBY: It is my suggestion to purchase a new, fresh SCOBY from a reputable source rather than using SCOBY from someone else. This way you are ensuring that you are starting with a fresh, pathogenic-free SCOBY. (See Resources, page 247.) Although part of the tradition of making kombucha is handing down SCOBY from one person to another (when making kombucha, the SCOBY then makes what is called a "baby" SCOBY—more to come on this), for safest practices, buy your own SCOBY to start fresh! Then reuse your own SCOBY.

Consume safe quantities of kombucha: Typically, consumption of 4 ounces per day is a safe place to start—and to stay; this is why most of the kombucha recipes in this book call for only ½ cup of kombucha, to help you keep your intake in check. Only start consuming kombucha if you are healthy. If you have health issues, discuss it with your health care provider before consuming kombucha.

Another reason to limit how much kombucha you consume is that, because of the brewing process, kombucha will have a small percentage of alcohol. Store-bought kombucha must have 0.5% alcohol or less. When making your own batch at home, it can vary. You can determine how much alcohol is in your kombucha with special equipment known as a hydrometer, which measures the specific gravity of the liquid. To determine the alcohol by volume (ABV), first measure the specific gravity of the liquid before you start and then measure the specific gravity of the finished kombucha. Then using an ABV calculator, you can determine the percentage of alcohol in your batch. Most homemade kombucha comes in around 0.5% or less for alcohol percentage.

original
KOMBUCHA

supplies

Kombucha home brewing kit that includes:

One 12-cup kombucha brewing jar

Temperature indicator strip, for the jar

Cotton cloth or cheesecloth (cover for the jar during brewing)

Rubber band

pH strips

Four 1-quart storage bottles or canning jars with lids

ingredients

12 cups water, divided

6 unflavored tea bags (see Note, page 27)

1 cup granulated sugar

SCOBY package (with SCOBY and liquid), or the reserved SCOBY and kombucha from a brewed batch (see details below)

My go-to source for kombucha brewing supplies is Kombucha Brooklyn (see Resources, page 247). There are also other sources, although some sell the SCOBY in a dehydrated state. The "live" SCOBY is easier to work with and produces a great batch of kombucha the first time, whereas it can take a few batches with a dehydrated SCOBY for productive brewing.

Before you start brewing your own kombucha, it is important to know that it is a process that takes time. Typically the brewing takes 7 to 14 days, although when you first start it could be 15 to 28 days, depending on brewing conditions and factors like temperature and water quality. **SERVES 48 (½ CUP EACH)**

preparing the tea and getting started

Make sure the work surface and your hands are clean.

In a saucepan, bring 4 cups of the water to a boil and add the tea bags. Remove from the heat and let the tea steep for 20 minutes. This step should not be completed in the brewing jar!

While your tea is brewing, wash your kombucha brewing jar with warm soapy water. Rinse thoroughly and dry.

Remove the tea bags from the pan. Add the sugar to the tea (this is food for the SCOBY), stirring to dissolve. Add the remaining 8 cups (½ gallon) cold water to the brewing jar, then add the brewed tea.

Place the temperature indicator strip on the side of your brewing bottle. Once the tea is below 90°F, it is time to add the SCOBY. Transfer the entire contents of your SCOBY package (liquid and SCOBY) to the brewing jar. Cover the top with the cotton cloth and the rubber band to hold it tight.

Move your brewing bottle to a warm, dark area (away from direct sunlight) to let the brewing begin. Ideally you want a temperature between 72°F and 80°F. Adjust the location to keep your jar within the range throughout the brewing process.

Days 1 to 2: Be patient! Within the first two days, a new "baby" SCOBY will form on the top of the jar and small bubbles will start to form. This is not mold but rather the beginning of the brewing process. Mold looks different than the SCOBY formation: Mold looks more like small white or black spots/specks on the top of the batch, versus the SCOBY which has more of a waxy look.

Days 3 to 4: A maturing culture will result in the growth and expansion of the bubbles, which will slowly take over the top of the jar.

Days 7 to 14: When to stop the brewing process is somewhat a matter of personal preference, depending on the taste profile you would like to achieve. Monitor the top of your jar: When the surface has been visibly taken over by the cultures (which should happen between day 7 and day 14), it is time to test the acidity of the tea with pH indicator strips. I like to bottle my batches when the pH is around 2.8 or 2.9, when it is slightly less acidic. The pH is an indicator of the sweetness: A pH of 3.1 is a sweeter flavor and a pH of 2.7 is more on the sour side. The longer the brewing process continues, the less sugar that is left in the tea (the SCOBY uses the sugar basically as food) and hence, a more sour taste and lower acidity.

As noted above, your first batch of kombucha may take longer to reach the point when the SCOBY has "taken over" the top of your brewing jar. Another indicator to monitor is the pH; if the pH is still above 3.0, let the brewing process continue to allow the SCOBY to consume more of the sugar, which creates a traditional kombucha flavor.

time to bottle

Once you have determined your kombucha is done fermenting, it is time to bottle your kombucha. Start by thoroughly washing your hands and remove the two SCOBYs (the one you added to the jar and the one that formed during fermentation, which typically separates from the original SCOBY and is referred to as the "baby"). Transfer the top SCOBY (the baby) and the top layer of the original SCOBY to a bowl and top with 1½ cups of your brewed kombucha; this becomes your SCOBY package for the next batch of kombucha. Discard the old SCOBY.

Sanitize your storage bottles or canning jars by washing them with warm soapy water or in the dishwasher. Fill your bottles with the remaining kombucha and cover, or get creative and flavor your kombucha (see the Variations, page 28).

Once your kombucha is bottled, let it sit at room temperature for 3 to 7 days to get bubbly, then transfer the bottles to the refrigerator to slow the fermentation process. Or, after bottling, transfer to the refrigerator for a non-bubbly version of kombucha.

choosing bottles

Many kombucha supply stores will have 32-ounce bottles to store your kombucha, but my personal pick is 1-quart canning jars. The reason I like the canning jar option is because one of my favorite ways to enjoy kombucha involves the second ferment and canning jars have a wide mouth that makes it easier to add things like orange slices to flavor the kombucha.

your next batch

Now it's time to brew your next batch of kombucha! You simply go back to the preparing the tea step. The difference is that this time you are going to use the baby SCOBY (newly developed) and the original SCOBY along with the 1½ cups of reserved kombucha as the SCOBY package for batch number two!

taking a break from making kombucha

If you don't want to immediately get another batch of kombucha brewing, you will need to store your SCOBY and reserved kombucha. Make your own SCOBY package by storing the SCOBY and the reserved 1½ cups kombucha in an airtight container in the refrigerator for up to several months. When you are ready to get started brewing again, follow the brewing process.

Note: Choosing tea for your kombucha: There is room to play around with the types of tea that you use to brew kombucha, although it is best to stay with pure tea varieties, like green tea and black tea, that do not have added flavors, as the flavors could disrupt the brewing process.

variations

THE SECOND FERMENT

After you make a batch of kombucha, you can naturally flavor it by doing what is called a second ferment. Basically, it means that you combine the original (aka plain) kombucha with fruits, spices, and/or herbs to further ferment the beverage and add natural flavors.

Below is a list of fermenting combinations to try, plus be creative and try your own flavor combinations!

Start with 4 cups kombucha in a glass jar or container and then add:

Black Cherry–Mint Kombucha: 12 fresh or frozen black cherries and 8 crushed mint leaves

Key Lime Kombucha: 4 fresh key limes, sliced into halves

Lemon Kombucha: 3 fresh lemon slices

Apple Kombucha: 3 apple slices

Blackberry Kombucha: 12 fresh blackberries

After adding your choice of fruit and/or herbs, cover the jar and set aside on the counter or in a cupboard, away from direct sunlight, for 12 to 24 hours. Strain out the fruit and chill the kombucha. Serve ½ cup of the flavored kombucha over ice.

---

nutrition facts* (per serving)

15 calories, 0 g fat, 0 g saturated fat, 0 g trans fat, 0 mg cholesterol, 5 mg sodium, 4 g carbohydrates, 0 g fiber, 1 g sugar, 0 g protein, 0% vitamin A, 0% vitamin C, 0% calcium, 0% iron

* The nutritional content will vary based on fruit combination and fermenting time. The above nutrition facts are based on plain, unflavored kombucha.

ginger
BEER

supplies

Two 1-quart glass jars with lids
Grater
Fine-mesh strainer

ingredients

4 ounces fresh ginger root, or more if needed

8 cups lukewarm water

1 cup granulated sugar

1½ tablespoons fresh lemon juice

¼ teaspoon active dry yeast

One of the key components of ginger beer is fresh ginger, which has natural healing properties like helping to settle an upset stomach. It has been found to have anti-inflammatory properties too, which can make it ideal for combating conditions like rheumatoid arthritis. Ginger beer will have a small amount of alcohol due to the fermentation process, although it is typically less than 0.5%, which is why commercial ginger beer is sold as a drink, not as a beer. SERVES 8 (1 CUP EACH)

directions

Sterilize the jars with boiling water or wash well with warm soapy water. Pat dry with a clean towel.

Peel the ginger, then, using a fine grater, grate into a coarse pulp. Position the strainer over a bowl and press the ginger root pulp into the strainer to release the juice. You'll need ¼ cup juice, so you may need to grate more ginger to get the right amount.

Divide the ginger juice and all of the remaining ingredients evenly between the 2 sterilized jars.

Cover each jar tightly with a lid and gently shake to dissolve the sugar. Store the sealed jars on the countertop (out of direct sunlight) for 24 to 36 hours. This will allow the beverage to ferment and for the carbonation to form. After 24 to 36 hours, check for carbonation by opening the jars and listening for the release of the carbonation and looking for the bubbles. Once the ginger beer is carbonated, store it in the refrigerator. The longer you let the beverage ferment, the bubblier it will be, so adjust the fermentation time according to your preference.

Store the bubbly ginger beer in the refrigerator for up to 7 days (this will slow the fermentation process) and enjoy the ginger beer just by itself or use it in one of the ginger beer recipes in this book.

nutrition facts (per serving)
100 calories, 0 g fat, 0 g saturated fat, 0 g trans fat, 0 mg cholesterol, 0 mg sodium, 25 g carbohydrates, 0 g fiber, 25 g sugar, 0 g protein, 0% vitamin A, 0% vitamin C, 0% calcium, 0% iron

traditional
PLAIN KEFIR

supplies

1-quart glass jar with a lid

Thermometer

Coffee filter or cheesecloth

Rubber band

ingredients

4 cups low-fat or nonfat milk (see Choosing Milk for Kefir, page 34)

¼ cup kefir grains or one 5-gram packet kefir starter (powder)

The history of kefir begins with the shepherds living on the slopes of the North Caucasus Mountains in Russia. The popularity of kefir spread throughout Eastern Europe and it is now widely consumed throughout much of the world. The people of the Caucasus region are known for their longevity and extraordinary health, attributed to their routine consumption of kefir. In fact, kefir was used as a treatment in hospitals in the former Soviet Union for conditions ranging from cancer to allergies to gastrointestinal disorders.

Making kefir requires kefir grains, which have a cauliflower-like texture and appearance and actually are not a grain at all. They are cultures of healthy bacteria and yeasts that are bound together in a matrix naturally created by the bacteria. The "grains" feed on sugar and produce lactic acid, alcohol, and carbon dioxide, resulting in a lightly carbonated dairy beverage. You will likely find the grains in a dehydrated state and so you need to rehydrate, or "activate," them (see the instructions on page 33). The best part is that kefir grains are reusable and can be used over and over again.

An alternative (and my favorite option) to kefir grains is kefir starter, which is a powdered freeze-dried mix. Each 5-gram packet of kefir starter makes one batch of kefir and is a single use.

Kefir grains and starters can be found via the Internet or in natural health food stores. For other suggestions, see Resources, page 247.

SERVES 4 (1 CUP EACH)

directions

If using dehydrated kefir grains, see the directions for rehydrating on the opposite page. Sterilize the 1-quart jar with boiling water or wash well with warm soapy water.

In a saucepan, heat the milk to 180°F. Place the pan of warm milk into an ice bath (a larger pan with cold water and ice cubes). Allow the mixture to cool until it reaches 80°F to 85°F.

Once the mixture is cooled, add the kefir grains or kefir starter. If using the starter, gently whisk it into the milk until well combined. Transfer the milk with the kefir grains or starter to the jar.

Cover the jar with a coffee filter or cheesecloth and secure with a rubber band. Let the mixture sit at room temperature for 12 to 24 hours to culture.

When the mixture has thickened, if you used kefir starter, cover the container with the lid and refrigerate. If you used kefir grains, the grains will have floated to the surface. Gently remove the grains with a wooden spoon and strain the kefir through a fine-mesh sieve. Return the freshly made kefir to a clean glass jar, cover, and refrigerate.

Storing the grains: After your batch of kefir is made, you can start another batch of kefir or you can dry the grains to store for later use. For short-term storage (up to 3 weeks): Store the kefir grains in 3 cups of milk in the refrigerator until you are ready to make another batch. For longer-term storage: Dry the grains, place on a clean paper towel, and air-dry for 3 to 5 days at room temperature. Store the grains in a storage container or bag in the refrigerator for up to 6 months.

nutrition facts* (per serving)

100 calories, 2½ g fat, 1½ g saturated fat, 0 g trans fat, 10 mg cholesterol, 105 mg sodium, 12 g carbohydrates, 0 g fiber, 12 g sugar, 8 g protein, 10% vitamin A, 0% vitamin C, 30% calcium, 0% iron

* Using 1% low-fat milk

rehydrating milk kefir grains

Starting with dehydrated kefir grains takes some patience, although once the grains are rehydrated, you can continue to make kefir and do not need to repeat this process.

Step 1: Fill a small container with 1 cup milk, then add the contents of the grain packet into the milk. Cover the container with cloth or a coffee filter and secure with a rubber band. Let the container sit on the counter or in a cupboard to culture for about 24 hours. Drain the grains (discard the milk) and start the process again.

Step 2: After you have completed step 1 two times, increase the milk to 2 cups and repeat the above process. You will know the kefir grains are activated once the milk is thickened within 24 hours and has a pleasant but sour smell. If the milk is still not thickened within 24 hours, repeat the process.

Step 3: Once the milk thickens within 24 hours, you know your kefir grains are rehydrated. You can now drink the milk and you can start making full batches of kefir.

choosing milk
for kefir

Any type of milk, from nonfat to whole, will work when making kefir. Choose what variety is your preference. A higher fat content will result in thicker kefir. If you can, choose a variety that is not UHT (ultra high temperature) processed for best results. UHT milk has a longer shelf life compared to traditionally pasteurized milk and *can* work for making kefir, but the best results are with milk that is not UHT treated.

Nondairy milk, including almond milk, hemp milk, coconut milk, etc., can be utilized as well, although especially with nut milks, the results can be inconsistent.

A word on whole milk: Two new studies recently published (in 2013) in the *European Journal of Nutrition* and *Scandinavian Journal of Primary Health Care* concluded that the consumption of whole milk does not contribute to obesity. In fact, over the 12-year period of the Scandinavian-based study, men who consumed whole milk, butter, and cream were less likely to become obese. This isn't the green light to go crazy with butter, whole milk, and cream, but it may suggest that at least for some, higher-fat milk products might be okay. Moving back to whole milk may or may not be right for you though. For example, if you have heart disease or a strong family history of heart disease, the extra saturated fat may put you at further risk. Stay tuned for more research developing in this area.

refrigerator
PLAIN KEFIR

supplies

One 1-quart glass jar with a lid

Thermometer

Coffee filter or cheesecloth

Rubber band

ingredients

4 cups low-fat or nonfat milk (see Choosing Milk for Kefir, opposite)

¼ cup kefir grains or one 5-gram packet kefir starter (powder)

Traditional kefir (page 31) is made by letting milk and kefir grains sit on the countertop to ferment. However, kefir can also be made in the refrigerator. This takes a few days longer than the countertop method, but still gets the job done! The basic difference: Refrigerator kefir needs to culture in the fridge for 5 days, while regular kefir needs just 1 day at room temperature. SERVES 4 (1 CUP EACH)

directions

Sterilize the jar with boiling water or wash well with warm soapy water.

In a saucepan, heat the milk to 180°F. Place the pan with the warm milk into an ice bath (a large pan filled with cold water and ice cubes). Allow the mixture to cool until it reaches 80° to 85°F.

Once the mixture is cooled, add the kefir grains or kefir starter. If using kefir starter, gently whisk it into the milk until well combined. Transfer the milk and kefir starter or grains to the jar.

Cover the jar with a coffee filter or cheesecloth and secure with a rubber band. Transfer the container to the refrigerator to let it brew for 5 days, gently shaking or stirring daily.

Once the mixture has thickened, if you used kefir starter cover the jar with the lid and refrigerate. If you used kefir grains, the kefir grains will have floated to the surface; gently remove the grains with a wooden spoon and strain the kefir through a fine sieve. Return the kefir to a clean glass jar, cover, and refrigerate.

See the tips for storing kefir grains between batches on page 32.

See the tips for storing kefir grains between batches on page 32.

nutrition facts* (per serving)
100 calories, 2½ g fat, 1½ g saturated fat, 0 g trans fat, 10 mg cholesterol, 105 mg sodium, 12 g carbohydrates, 0 g fiber, 12 g sugar, 8 g protein, 10% vitamin A, 0% vitamin C, 30% calcium, 0% iron

*Using 1% low-fat milk

water kefir

supplies

Two 1-quart glass jars

Thermometer

Cheesecloth or coffee filter

Rubber band

Fine-mesh strainer

ingredients

4 cups water

¼ cup granulated sugar

¼ cup water kefir grains

Water kefir uses a different type of kefir grains, known as water kefir grains or sugar kefir grains. These grains have a different look to them, almost a crystal-like texture, and are rich in healthy bacteria. Once water kefir grains are combined with sugar water, a fermented beverage develops. It is an excellent way to have a nondairy fermented beverage, and its light fizzy texture and subtly sweet taste makes it an optimal replacement for soda.

Water kefir grains can be found on the Internet or in some natural food stores and are typically packaged in a dehydrated state. Follow the package rehydration instructions that come with your grains (or see Directions on page 33) before your first use. Water kefir grains can be reused like milk kefir grains. (See Resources, page 247, for sources of kefir grains.) **SERVES 4 (1 CUP EACH)**

directions

If using dehydrated water kefir grains, see the directions for rehydrating on page 38. Sterilize the jars with boiling water or wash them well with warm soapy water.

Add the water to a saucepan and heat to at least 180°F (the water will begin to lightly bubble). Add the sugar and stir until it is completely dissolved.

Transfer the sugar water to one of the prepared jars and let the water cool to 68° to 85°F. Add the kefir grains. Cover the jar with cheesecloth or a coffee filter and use a rubber band to secure the covering. Place the jar in a warm spot (68° to 85°F) and let it sit for 24 to 48 hours to culture.

Strain the cultured water kefir using the strainer. Transfer the cultured liquid to the other prepared jar, cover with a lid, and store in the refrigerator for up to 7 days, or consume!

You can then start another batch of water kefir by following the above steps, or store the grains for later use (see Storing Water Kefir Grains, opposite).

---

nutrition facts* (per serving)
15 calories, 0 g fat, 0 g saturated fat, 0 g trans fat, 0 mg cholesterol, 0 mg sodium, 4 g carbohydrates, 0 g fiber, 4 g sugar, 0 g protein, 0% vitamin A, 0% vitamin C, 0% calcium, 0% iron

* It is difficult to calculate the nutrition facts for water kefir because the nutrition varies based on how much sugar is fermented by the kefir grains during the process.

rehydrating water kefir grains

Starting with water kefir grains in the dehydrated form takes some patience, although it is slightly more simple compared to rehydrating milk grains. Once the water kefir grains are rehydrated, you can continue to make kefir and do not need to repeat this process.

Step 1: Warm 3 cups water and then stir in ¼ cup sugar and dissolve. Cool the water to 68° to 85°F and transfer to a storage container. Add the packet of dehydrated water kefir grains to the sugar water. Cover the container with a cloth or coffee filter and secure with a rubber band.

Step 2: Place the container on a counter or in a cupboard and let sit for 3 to 5 days to allow the grains to rehydrate. The water kefir grains will have a translucent and plump look to them, which indicates that they are ready to be used for your first batch. Drain, and discard the sugar water. Use the grains to make your first batch by following the recipe.

storing water kefir grains

You can refrigerate water kefir grains in a sugar solution (4 cups water with ¼ cup sugar) for up to 3 weeks. To store for a longer time, dry the grains by placing on a paper towel and letting them sit at room temperature for 3 to 5 days. Once dried, store in a storage bag or container in the refrigerator for up to 6 months.

coconut water
KEFIR

supplies
Two 1-quart glass jars with lids
Thermometer
Cheesecloth or coffee filter
Rubber band
Fine-mesh strainer

ingredients
4 cups coconut water (see Note)
1 tablespoon granulated sugar
¼ cup water kefir grains

Coconut water is the base for this water kefir. The steps are the same as making water kefir. The primary difference is that due to the naturally occurring sugar in coconut water, the amount of added sugar is reduced. SERVES 4 (1 CUP EACH)

directions
Sterilize the jars with boiling water or wash them well with warm soapy water.

Add the coconut water to a saucepan and heat to at least 180°F (the water will begin to lightly bubble). Add the sugar and stir until it is completely dissolved.

Transfer the sugar–coconut water to one of the prepared jars and let cool to 68° to 85°F. Once the water is cooled, add the kefir grains. Cover the jar with cheesecloth or a coffee filter and use a rubber band to secure it.

Place the jar in a warm spot (68° to 85°F) and let it sit for 24 to 48 hours to culture.

Strain the cultured water kefir using the strainer. Transfer the cultured liquid to the other prepared jar, cover with a lid, and store in the refrigerator for up to 7 days (see Storing Water Kefir Grains on page 39) or consume!

Note: Use store-bought plain coconut water or try Coco Hydro instant coconut water—follow the package instructions to mix the powder with water to make coconut water.

nutrition facts* (per serving)
5 calories, 0 g fat, 0 g saturated fat, 0 g trans fat, 0 mg cholesterol, 10 mg sodium, 7 g carbohydrates, 0 g fiber, 6 g sugar, 0 g protein, 0% vitamin A, 25% vitamin C, 0% calcium, 0% iron

* It is difficult to calculate the nutrition facts for coconut water kefir because the nutrition varies based on how much sugar is fermented by the kefir grains during the process.

fruity water
KEFIR

supplies

Two 1-quart glass jars

Thermometer

Cheesecloth or coffee filter

Rubber band

Fine-mesh strainer

ingredients

3½ cups water

¼ cup granulated sugar

½ cup 100 percent fruit juice (such as apple or grape)

¼ cup water kefir grains

This nondairy water kefir with fruit juice makes a delicious juice-flavored, probiotic-rich beverage. Plus you gain the boost from the vitamins and minerals from the fruit juice. SERVES 4 (1 CUP EACH)

directions

Sterilize the jars with boiling water or wash them well with warm soapy water.

Add the water to a saucepan and heat to at least 180°F. Add the sugar and stir until completely dissolved. Stir in the fruit juice.

Transfer the juice-water mixture to one of the prepared jars and let cool to 68° to 85°F. Once the water is cooled, add the water kefir grains and cover the jar with cheesecloth or a coffee filter and use a rubber band to secure it.

Place the jar in a warm spot (68° to 85°F) and let it sit for 24 to 48 hours for the beverage to ferment. Note: The longer you leave it, the more the sugar ferments, which will lower the carbohydrate content of the drink.

Strain the cultured water kefir using the strainer. Transfer the cultured liquid to the other prepared jar, cover with a lid, and store in the refrigerator for up to 7 days, or consume!

See tips for storing the grains on page 39.

See tips for storing the grains on page 39.

nutrition facts* (per serving)

45 calories, 0 g fat, 0 g saturated fat, 0 g trans fat, 0 mg cholesterol, 0 mg sodium, 12 g carbohydrates, 0 g fiber, 12 g sugar, 0 g protein, 0% vitamin A, 0% calcium, 2% vitamin C, 0% iron

* It is difficult to calculate the nutrition facts for fruity water kefir because the nutrition varies based on how much sugar is fermented by the kefir grains during the process.

homemade
YOGURT

supplies

One 1-quart glass jar with a lid

Thermometer

Hand towel

String

ingredients

4 cups low-fat or nonfat milk

¼ cup (1 packet) yogurt starter or 5 ounces store-bought low-fat or nonfat plain yogurt (see Note)

Making your own probiotic-rich yogurt is actually pretty simple! The key is having yogurt starter. You can use store-bought yogurt as the starter, or you can purchase yogurt starter, which is live and active cultures in a freeze-dried powder, in packets or jars.

Tip: When buying cow's milk, opt for varieties that are not UHT (ultra high temperature) processed; they tend to culture better than UHT varieties. SERVES 4 (1 CUP EACH)

directions

Sterilize the 1-quart jar with boiling water or wash very well with warm soapy water.

In a small saucepan, heat the milk to 180°F. Remove from the heat and allow to cool until it reaches a temperature between 105°F and 115°F.

Add the yogurt starter and mix well. Transfer the mixture to the sterilized jar and cover. Wrap the jar of yogurt in a hand towel and tie with string to secure; this is to help maintain a temperature of 110°F to 112°F—the culturing temperature zone (see Maintaining the Culturing Temperature Zone, directions on page 46)—which allows the cultures to form.

Allow the yogurt to sit for 6 to 12 hours for the yogurt to thicken.

Once the curd forms and the mixture has thickened, refrigerate the yogurt overnight or for at least 4 hours to let the mixture set. Serve or store in the refrigerator for up to 7 days. Stir before use.

Note: If using store-bought plain yogurt as the starter, choose one that says on the container "live and active cultures" or "active cultures," as the cultures are what will make your yogurt.

variation
If you prefer the texture of Greek yogurt, you can transform your homemade yogurt into Greek yogurt by using a Greek yogurt maker (see Resources on page 247, for details).

nutrition facts* (per serving)
100 calories, 2½ g fat, 1½ g saturated fat, 0 g trans fat, 10 mg cholesterol, 105 mg sodium, 12 g carbohydrates, 0 g fiber, 12 g sugar, 8 g protein, 10% vitamin A, 0% vitamin C, 30% calcium, 0% iron

*Using 1% low-fat milk

maintaining
the culturing
temperature zone

There are many ways you can keep your milk in the right temperature zone to culture; here are a couple of DIY options:

- Place your wrapped jar on a heating pad and then cover again with another towel; this will help to keep the mixture in the culturing temperature zone.
- Fill your slow cooker with enough water to mostly cover your jar of milk and cultures. Heat the water to 105° to 112°F, then turn the cooker off and check the temperature periodically; you may need to turn the heat back to low to keep the water in the culturing zone.
- Using a yogurt maker allows for the most consistency and ease when making yogurt and takes the work out of maintaining the proper temperature. There are a variety of yogurt maker options available at a range of price points; see Resources on page 247.

nondairy
YOGURT

supplies

One 1-quart glass jar with a lid

Thermometer

Hand towel

String

ingredients

4 cups nondairy milk (coconut milk, almond milk, soy milk, or rice milk; see Note, page 48)

¼ cup (1 packet) nondairy yogurt starter or 5 ounces store-bought nondairy yogurt (see Note, page 48)

Making your own probiotic-rich nondairy yogurt can save you a bundle! Each single serving of nondairy yogurt from the health food market can cost $1.50 to $2.00 — but you can make a whole batch of nondairy yogurt for only around $2.00.

One key is having nondairy yogurt starter or store-bought non-dairy yogurt (such as almond milk or coconut milk) as the starter. Nondairy yogurt starter is found in a powdered form, in packets. The yogurt starter contains healthy bacteria (such as *L. acidophilus*) that ferments the milk and makes the yogurt. **SERVES 4 (1 CUP EACH)**

directions

Sterilize the 1-quart jar with boiling water or wash very well with warm soapy water.

In a saucepan, heat the milk to 180°F. Remove from the heat and allow the mixture to cool to 105° to 115°F, 5 to 10 minutes.

Stir in the yogurt starter and mix well. Transfer to the clean glass jar and cover. Wrap the jar in a hand towel and tie with a string to secure; this is to help maintain a temperature of 110° to 112°F (the culturing temperature zone) for the cultures to form. (See Maintaining the Culturing Temperature Zone, opposite, for DIY yogurt maker ideas and tips.)

Allow the yogurt to sit for 6 to 12 hours for the yogurt to thicken. Once the curd forms and thickens, refrigerate the yogurt overnight or for at least 4 hours to let the mixture set. Serve, or store in the refrigerator for up to 7 days. Stir before use.

Note: Nondairy milk tends to produce a thinner yogurt compared to traditional homemade yogurt. To thicken the yogurt, you can stir in dry milk powder, pea protein powder, or use a thickener like arrowroot.

Note: If using store-bought plain yogurt as the starter, choose one that says on the container "live and active cultures" or "active cultures," as the cultures are what will make your yogurt.

nutrition facts (per serving)
80 calories, 4 g fat, 1 g saturated fat, 0 g trans fat, 0 mg cholesterol, 140 mg sodium, 5 g carbohydrates, 1 g fiber, 1 g sugar, 7 g protein, 6% vitamin A, 0% vitamin C, 30% calcium, 6% iron

* Using unsweetened soy milk

homemade
ALMOND MILK

supplies

One 1-quart glass jar with a lid

Food processor or blender

Strainer

Cheesecloth

ingredients

1 cup raw organic almonds

2 cups water, plus more for soaking

2 teaspoons pure maple syrup or agave nectar, or to taste

Almond milk has been found to naturally have small amounts of probiotic properties and is a great stand-in for cow's milk for those that choose to follow a vegan or Paleo diet. Plus, it works as a great base for making nondairy probiotic-rich yogurt (see page 47).

A note about vitamin D: If you are going to start making most of your own beverages like almond milk, you need to carefully consider how much vitamin D you are getting from other food sources (such as cod liver oil, tuna, and salmon), as there are very few foods that naturally have vitamin D occurring in them. The daily recommendation for an adult's vitamin D intake is 600 IU. Most of the vitamin D that we consume is in the form of supplemental vitamin D from fortified products like store-bought almond milk, cow's milk, and yogurt. So consider taking a supplement if you are going to consume mostly homemade almond milk. Additionally, store-bought almond milk is fortified with calcium, another important mineral. Make sure you are getting enough calcium from other foods (like yogurt, cheese, broccoli, kale, and tofu) if you are making your own almond milk.

SERVES 2 (1 CUP EACH)

directions

Sterilize the 1-quart jar with boiling water or wash very well with warm soapy water.

Place the almonds in a bowl, completely cover with water, and cover the bowl with a tight-fitting lid. Refrigerate overnight or for up to 2 days.

Drain the almonds from the soaking water and rinse under cool running water. Place the drained almonds in food processor or blender and add the 2 cups water. Blend on high for 2 to 4 minutes, until the water is white and opaque.

Line a strainer with cheesecloth and place over a container or bowl. Pour the almond milk mixture into the strainer. Gather the sides of the cheesecloth and squeeze to remove all the liquid. Discard the solids.

Stir the maple syrup into the milk and adjust the amount of sugar to your personal preference.

Store the almond milk in a covered glass jar in the refrigerator for up to 2 days.

nutrition facts* (per serving)

40 calories, 2 g fat, 0 g saturated fat, 0 g trans fat, 0 mg cholesterol, 0 mg sodium, 5 g carbohydrates, 0 g fiber, 4 g sugar, <1 g protein, 10% vitamin A, 0% vitamin C, 0% calcium, 2% iron

*Nutrition facts are approximate and can vary slightly from batch to batch.

vanilla
LASSI

1½ cups store-bought low-fat or nonfat plain yogurt or Homemade Yogurt (page 44)

½ cup low-fat or nonfat milk

2 teaspoons granulated sugar (optional)

1 teaspoon vanilla extract

Lassi, a yogurt-based drink popular in India and Pakistan, is a blend of yogurt, water, spices, and sometimes fruit. Traditionally a lassi is savory and includes cumin. Sometimes they are sweetened with fruit instead of spices. For a fruity lassi, try Larry's Mango Lassi on page 69. SERVES 2 (1 CUP EACH)

directions

Combine the yogurt and milk in a mixing bowl and whisk until combined. Stir in the sugar, if desired, and vanilla extract. Divide among glasses and serve or store in the refrigerator for up to 3 to 4 days.

nutrition facts (per serving)

130 calories, 3 g fat, 2 g saturated fat, 0 g trans fat, 15 mg cholesterol, 140 mg sodium, 15 g carbohydrates, 0 g fiber, 15 g sugar, 10 g protein, 2% vitamin A, 35% calcium, 4% vitamin C, 0% iron

SMOOTHIE
AND BEVERAGE
RECIPES

IMMUNE-BOOSTING

When it comes to boosting your immune system, there are a couple of important vitamins and minerals that play key roles, including vitamin A, vitamin C, and zinc. Food sources of each of these that are included into the recipes in this section include:

Vitamin A: red peppers, mango, carrots, cantaloupe, milk, and yogurt

Vitamin C: red peppers, oranges and orange juice, strawberries, kiwi, and cantaloupe

Zinc: yogurt and milk

Try having one of the smoothies daily during cold and flu season to help keep your immune system working at its peak.

avocado orange
SMOOTHIE

1 cup light almond milk or Homemade Almond Milk (page 49)

2 ounces tempeh

1 orange, peeled and seeded

½ avocado, pitted and peeled

½ cup packed baby spinach

1 teaspoon agave nectar

This smoothie is rich in vitamin C, a water-soluble vitamin that is an important antioxidant. In addition, each serving provides 25 percent of your daily requirement of vitamin A, an important fat-soluble vitamin (which means it requires fat to be utilized by the body). The good news is that the drink also has 9 grams of fat, coming mostly from the avocado, to help absorb the vitamin A. SERVES 2 (1 CUP EACH)

directions

Combine all of the ingredients in a blender (see Note) and blend until smooth.

Divide between 2 glasses and serve immediately.

Note: This recipe works best with a high-speed blender. If using a traditional kitchen blender, chop the orange and tempeh before adding to the blender.

nutrition facts (per serving)
210 calories, 9 g fat, 2 g saturated fat, 0 g trans fat, 5 mg cholesterol, 75 mg sodium, 26 mg carbohydrates, 5 g fiber, 16 g sugar, 9 g protein, 25% vitamin A, 70% vitamin C, 20% calcium, 6% iron

blueberry
CINNAMON
CRUSH

1 cup low-fat or nonfat plain Greek yogurt

½ cup low-fat or nonfat milk

1 cup frozen blueberries

1 tablespoon agave nectar

½ teaspoon ground cinnamon

4 or 5 ice cubes

This smoothie gets its immune-boosting properties from deep dark blueberries, which are rich in anthocyanins, a powerful antioxidant. Plus it boasts an additional immune boost from the zinc that is naturally occurring in the milk; each cup of milk has about 1 milligram of zinc. It also has cinnamon, which for centuries has been used by cultures to treat gastrointestinal problems and other ailments. Although modern science has not yet been able to prove the health connection, the warm cinnamon flavor complements the yogurt and blueberries quite well. SERVES 2 (1 CUP EACH)

directions

Combine the yogurt and milk in a blender and blend until smooth.

Add the blueberries, agave, cinnamon, and ice cubes. Blend until icy and smooth.

Divide between 2 glasses and serve, or store in the refrigerator for up to 3 or 4 days.

Note: It is best to enjoy smoothies immediately after making, although they can be saved in the refrigerator for up to 3 or 4 days. Before serving, stir the smoothie well.

nutrition facts (per serving)
180 calories, 1 g fat, 1 g saturated fat, 0 g trans fat, <5 mg cholesterol,
80 mg sodium, 29 g carbohydrates, 3 g fiber, 25 g sugar, 15 g
protein, 4% vitamin A, 10% vitamin C, 20% calcium, 2% iron

carrot
KOMBUCHA

½ apple, peeled and cored

½ cup 100 percent carrot juice (such as from Bolthouse Farms)

3 fresh strawberries, hulled

2 ice cubes

1 cup Ginger Kombucha (page 155) or GT's Enlightened Organic Raw Gingerade Kombucha

Thanks to the carrot juice, each serving of this bright orange smoothie has 180 percent (8,774 IU) of the daily value for vitamin A, which plays crucial roles in the body including helping your skin and eyes. When shopping for carrot juice, select 100 percent carrot juice for the maximum nutrition benefit. **SERVES 2 (1 CUP EACH)**

directions

Combine the apple, carrot juice, strawberries, and ice cubes in a blender and blend until smooth.

Divide the carrot juice mixture between 2 glasses (about ½ cup each). Stir ½ cup of the kombucha into each glass. Serve, or store in the refrigerator for up to 3 or 4 days.

nutrition facts (per serving)
60 calories, 0 g fat, 0 g saturated fat, 0 g trans fat, 0 mg cholesterol, 50 mg sodium, 14 g carbohydrates, 2 g fiber, 10 g sugar, <1 g protein, 180% vitamin A, 2% calcium, 30% vitamin C, 2% iron

owen's orange
CRUSH

1 orange, peeled and seeded

1 cup 100 percent orange juice

½ cup low-fat or nonfat plain Greek yogurt

4 or 5 ice cubes

Tons (well, a couple hundred to be more exact) of smoothies were prepared in preparation for this book. My faithful panel of taste testers eagerly tried nearly all of the recipes over the course of the tests, and they each picked a favorite. This smoothie is our oldest son's favorite, and I hope that you will enjoy it as much as he does! Each serving has 180 percent of the immune-boosting vitamin C you need every day. Plus the Greek yogurt provides extra immune-boosting power with its naturally occurring zinc. **SERVES 2 (1 CUP EACH)**

directions

Combine all of the ingredients in a blender (see Note) and blend until frothy and smooth. Serve immediately or refrigerate for up to 3 or 4 days.

Note: This works best with a high-speed blender. If using a kitchen blender, chop the oranges into smaller pieces first before blending.

nutrition facts (per serving)
120 calories, ½ g fat, 0 g saturated fat, 0 g trans fat, 0 mg cholesterol, 30 mg sodium, 22 g carbohydrates, 2 g fiber, 13 g sugar, 7 g protein, 4% vitamin A, 10% calcium, 180% vitamin C, 2% iron

carrot immune
BOOST

½ apple, peeled and cored

½ cup 100 percent carrot juice (such as from Bolthouse Farms)

½ cup low-fat or nonfat vanilla bean yogurt

3 fresh strawberries, hulled

2 ice cubes

The creaminess of the yogurt and the carrot juice combined with the fresh flavors from the strawberries and apple make for a perfect immune boost. This smoothie provides 360 percent of the daily value for vitamin A (18,051 IU) and 100 percent for vitamin C (60 milligrams). SERVES 1 (1 CUP)

directions

Combine the apple, carrot juice, yogurt, strawberries, and ice cubes in a blender (see Note) and blend until smooth.

Serve, or store in the refrigerator for up to 3 or 4 days.

Note: This works best with a high-speed blender. If using a kitchen blender, chop the apple into smaller pieces before blending.

nutrition facts (per serving)

220 calories, 2½ g fat, 1½ g saturated fat, 0 g trans fat, 15 mg cholesterol, 140 mg sodium, 41 g carbohydrates, 4 g fiber, 32 g sugar, 10 g protein, 360% vitamin A, 25% calcium, 100% vitamin C, 6% iron

fresh mango
KEFIR

1½ cups cubed fresh mango (from 2 mangos)

1 cup plain kefir or Traditional Plain Kefir (page 31)

½ cup light almond milk or Homemade Almond Milk (page 49)

¼ cup 100 percent carrot juice (such as from Bolthouse Farms)

Fresh fruit whips up with the creaminess of kefir—the result is a delicious nutrient-packed drink. Making your own flavored kefir is a great way to include your own taste preferences. This kefir gets its immune-boosting properties from fresh mango and carrot juice. Each serving has 130 percent of the daily value for vitamin A (6,333 IU) and 100 percent for vitamin C (60 milligrams); both are important nutrients when it comes to boosting immune response. SERVES 2 (1 CUP EACH)

directions

Combine all of the ingredients in a blender and blend until frothy and smooth.

Divide between 2 glasses and serve, or store in the refrigerator for up to 3 or 4 days.

nutrition facts (per serving)
210 calories, 2½ g fat, 1 g saturated fat, 0 g trans fat, 0 mg cholesterol, 135 mg sodium, 43 g carbohydrates, 4 g fiber, 38 g sugar, 7 g protein, 130% vitamin A, 100% vitamin C, 30% calcium, 2% iron

kiwi strawberry
SMOOTHIE

½ cup low-fat or nonfat plain Greek yogurt

½ cup light almond milk or Homemade Almond Milk (page 49)

2 fresh kiwis, peeled

4 to 5 fresh strawberries, hulled

¼ avocado, pitted and peeled

2 teaspoons honey (optional)

Boost your immune system with this smoothie, which provides 180 percent (107 milligrams) of the vitamin C you need every day, plus 46 percent of the vitamin K (36 micrograms). Vitamin K helps your body in many ways, including making proteins for healthy bones and tissues. **SERVES 2 (1 CUP EACH)**

directions

Combine all of the ingredients in a blender and blend until smooth.

Divide between 2 glasses and serve, or store in the refrigerator for up to 3 or 4 days.

nutrition facts (per serving)
170 calories, 4½ g fat, 0 g saturated fat, 0 g trans fat, 0 mg cholesterol, 75 mg sodium, 26 g carbohydrates, 5 g fiber, 18 g sugar, 8 g protein, 4% vitamin A, 180% vitamin C, 20% calcium, 4% iron

mango blue
SMOOTHIE

1 cup frozen blueberries
1 cup frozen mango slices
1 cup low-fat or nonfat milk

¾ cup low-fat or nonfat plain Greek yogurt
1 tablespoon honey (or less)

The flavor combination of mango and blueberry is delicious! Adjust the amount of honey according to your taste preference and the sweetness of the fruit. SERVES 2 (1½ CUPS EACH)

directions

Combine all of the ingredients except the honey in a blender and blend until smooth. Sweeten to taste with the honey.

Divide between 2 glasses and serve, or store in the refrigerator for up to 3 or 4 days.

nutrition facts (per serving)
230 calories, 1½ g fat, 1 g saturated fat, 0 g trans fat, 5 mg cholesterol, 95 mg sodium, 43 g carbohydrates, 3 g fiber, 38 g sugar, 14 g protein, 20% vitamin A, 50% vitamin C, 25% calcium, 2% iron

mango banana
SMOOTHIE

1 banana

1 cup frozen cubed mango

½ cup low-fat or nonfat milk

½ cup Vanilla Bean Smoothie (page 128) or vanilla bean yogurt

The health benefits from mangoes extend beyond fiber and vitamin C. The beautiful yellow-orange color contributes to the antioxidant powder of the fruit; the pigment is part of a family called carotenoids, which are powerful antioxidants. SERVES 2 (1 CUP EACH)

directions

Combine all of the ingredients in a blender and blend until smooth and icy.

Divide between 2 glasses and serve, or store in the refrigerator for up to 3 or 4 days.

nutrition facts (per serving)

160 calories, 1 g fat, ½ g saturated fat, 0 g trans fat, <5 mg cholesterol, 50 mg sodium, 34 g carbohydrates, 3 g fiber, 26 g sugar, 6 g protein, 15% vitamin A, 15% calcium, 45% vitamin C, 2% iron

larry's mango
LASSI

1 cup Vanilla Lassi (page 52) or store-bought vanilla lassi

1 cup frozen cubed mango

This twist on the traditional savory nature of lassi uses vanilla lassi as a base. You can use my lassi recipe, or pick up some bottled vanilla lassi from the grocery store. The recipe gets its name from my youngest taste tester: Our son Larry *loves* mango and pronounced this recipe his favorite of the book. SERVES 1 (1½ CUPS)

directions

Combine the lassi and mango in a blender and blend until smooth. Serve, or store in the refrigerator for up to 3 or 4 days.

nutrition facts (per serving)
250 calories, 3½ g fat, 2 g saturated fat, 0 g trans fat, 15 mg cholesterol, 135 mg sodium, 48 g carbohydrates, 5 g fiber, 42 g sugar, 9 g protein, 25% vitamin A, 30% calcium, 80% vitamin C, 2% iron

orange
KOMBUCHA

1 cup Original Kombucha (page 23) or GT's Enlightened Organic Raw Original Kombucha

1 orange, cut into 5 or 6 slices

This is hands down my favorite way to enjoy kombucha. The natural flavor of the oranges helps to mellow the strong kombucha, plus the orange boosts the vitamin C to 45 percent (28 milligrams) of your daily requirement. Try combining ½ cup of the orange kombucha with naturally flavored orange seltzer water—it is so refreshing.

SERVES 1 (1 CUP)

directions

Combine the kombucha in a glass jar or container with 4 of the orange slices. Cover tightly and let the mixture set at room temperature for 6 to 12 hours, depending on the desired flavor intensity. Once the kombucha is flavored, store in the refrigerator for up to 3 or 4 days.

Strain the kombucha and serve over ice, garnishing with a fresh orange slice or two.

nutrition facts (per serving)
60 calories, 0 g fat, 0 g saturated fat, 0 g trans fat, 0 mg cholesterol, 10 mg sodium, 13 g carbohydrates, 2 g fiber, 2 g sugar, <1 g protein, 2% vitamin A, 2% calcium, 45% vitamin C, 2% iron

orange power
SMOOTHIE

1 cup light almond milk or
Homemade Almond Milk (page 49)

1 cup cubed fresh cantaloupe

1 cup cubed fresh papaya

1 cup frozen cubed butternut squash

¾ cup plain cultured almond milk
yogurt or Nondairy Yogurt (page 47)

½ cup cubed fresh mango

2 teaspoons honey (optional)

Providing more than 100 percent of your daily requirements of vitamins A and C, this rich, orange-colored smoothie is a powerful immune booster. Researchers from the Centers for Disease Control and Prevention found that those with the highest levels of alpha-carotene (found in orange fruits and vegetables) were 40 percent less likely to have died from any cause than those with the lowest levels or none in their bloodstream. SERVES 3 (ABOUT 1 CUP EACH)

directions

Combine all of the ingredients in a blender and blend until icy and smooth.

Divide among 3 glasses and serve, or store in the refrigerator for up to 3 or 4 days.

nutrition facts (per serving)
160 calories, 2 g fat, 0 g saturated fat, 0 g trans fat, 0 mg cholesterol, 85 mg sodium, 35 g carbohydrates, 4 g fiber, 22 g sugar, 2g protein, 160% vitamin A, 25% calcium, 120% vitamin C, 4% iron

orange
STRAWBERRY
FREEZE

1 cup low-fat or nonfat plain Greek yogurt

1 cup frozen strawberries

½ cup 100 percent orange juice

½ orange, peeled and seeded

1 tablespoon honey

4 or 5 ice cubes

This refreshing drink provides 180 percent (107 milligrams) of the daily value of vitamin C, thanks to the mix of strawberries, orange, and orange juice! Vitamin C, an important antioxidant, has been shown to help regenerate other antioxidants in the body. In addition, it plays an immune-boosting role and aids in the absorption of iron, specifically non-heme iron, which comes from plant sources.

SERVES 2 (1 CUP EACH)

directions

Combine the yogurt, frozen strawberries, and orange juice in a blender (see Note) and blend until smooth. Add the orange, honey, and ice cubes and blend until the mixture has a smoothie texture.

Divide between 2 glasses and serve, or store in the refrigerator for up to 3 or 4 days.

Note: This works best with a high-speed blender. If using a standard kitchen blender, first chop the orange into small pieces.

nutrition facts (per serving)

170 calories, 0 g fat, 0 g saturated fat, 0 g trans fat, 0 mg cholesterol, 55 mg sodium, 31 g carbohydrates, 3 g fiber, 26 g sugar, 13 g protein, 2% vitamin A, 180% vitamin C, 15% calcium, 2% iron

peach
KEFIR

1 cup frozen sliced peaches (plain, without syrup)

½ cup plain kefir or Traditional Plain Kefir (page 31)

½ cup 100 percent orange juice

1 teaspoon agave nectar (optional)

This peach kefir is another avenue for packing in powerful alpha-carotene, which is linked to reducing the likelihood of dying of cancer or heart disease. Plus, each serving has 80 percent (40 milligrams) of the daily value for vitamin C. SERVES 2 (1 CUP EACH)

directions

Combine the frozen peaches, kefir, and orange juice in a blender and blend until smooth. Taste, and add agave nectar if needed.

Divide between 2 glasses and serve, or store in the refrigerator for up to 3 or 4 days.

nutrition facts (per serving)

90 calories, 1 g fat, 0 g saturated fat, 0 g trans fat, 0 mg cholesterol, 30 mg sodium, 17 g carbohydrates, 1 g fiber, 15 g sugar, 4 g protein, 8% vitamin A, 80% vitamin C, 8% calcium, 2% iron

roasted red
PEPPER JUICE

1 cup Orange Power Smoothie (page 71)

1 cup jarred roasted red peppers, drained

1 teaspoon organic mellow white miso

¼ teaspoon garlic powder

If you feel a cold coming on or everyone around you is sneezing, add this recipe to your routine to give your immune system a boost! Each serving provides more than twice the amount of vitamin A you need each day, and four times the amount of vitamin C. **SERVES 2 (1 CUP EACH)**

directions

Combine all of the ingredients in a blender and blend until smooth.

Divide between 2 glasses and serve, or store in the refrigerator for up to 3 or 4 days.

nutrition facts (per serving)
220 calories, 2 g fat, 0 g saturated fat, 0 g trans fat, 0 mg cholesterol, 425 mg sodium, 58 g carbohydrates, 5 g fiber, 23 g sugar, 3 g protein, 230% vitamin A, 25% calcium, 450% vitamin C, 6% iron

fresh
STRAWBERRY
SMOOTHIE

8 fresh strawberries, hulled

1 cup low-fat or nonfat vanilla bean Greek yogurt

½ cup light almond milk or Homemade Almond Milk (page 49)

4 or 5 ice cubes (optional)

Strawberries are another vitamin C–packed fruit! Each serving of this smoothie has 120 percent (70 milligrams) of the vitamin C you need each day. Plus, it tastes delicious and has a creamy texture. For more of an icy texture, add the optional ice cubes. SERVES 2 (1 CUP EACH)

directions

Combine all of the ingredients in a blender and blend until smooth.

Divide between 2 glasses and serve, or store in the refrigerator for up to 3 or 4 days.

nutrition facts (per serving)
160 calories, 3 g fat, 1½ g saturated fat, 0 g trans fat, 15 mg cholesterol, 105 mg sodium, 26 g carbohydrates, 2 g fiber, 19 g sugar, 9 g protein, 15% vitamin A, 35% calcium, 120% vitamin C, 6% iron

tangerine
FREEZE

2 fresh tangerines, peeled and seeded

½ cup plain kefir or Traditional Plain Kefir (page 31)

½ cup coconut milk beverage (such as from So Delicious)

1 teaspoon coconut palm sugar (optional)

4 or 5 ice cubes

This smoothie's sweet flavor comes from tangerines, its creaminess from kefir, and its hint of coconut from coconut milk. The unique, slightly nutty flavor of coconut palm sugar works well here, but a word of caution: While it has been found to be higher in antioxidants compared to other types of sugar, coconut palm sugar is still sugar and the quantity still needs to be limited! SERVES 2 (1½ CUPS EACH)

directions

Combine the tangerines, kefir, and coconut milk in a blender and blend until smooth. Add the sugar, if desired, and ice cubes. Blend until frothy.

Divide between 2 glasses and serve, or store in the refrigerator for up to 3 or 4 days.

nutrition facts (per serving)

120 calories, 1 g fat, ½ g saturated fat, 0 g trans fat, 0 mg cholesterol, 45 mg sodium, 25 g carbohydrates, 3 g fiber, 21 g sugar, 5 g protein, 25% vitamin A, 15% calcium, 70% vitamin C, 2% iron

tropical dream
SMOOTHIE

1 cup frozen cubed pineapple

½ cup low-fat or nonfat plain yogurt or Homemade Yogurt (page 44)

½ cup light almond milk or Homemade Almond Milk (page 49)

½ cup 100 percent orange juice

1 tablespoon shredded coconut

1 teaspoon agave nectar (optional)

The tropical combination of pineapple, orange juice, and coconut makes for a delicious smoothie that can boost your immune system! Each serving has 110 percent of the daily value for Vitamin C (67 milligrams) to provide an immune boost. The smoothie makes incredible ice pops as well—freeze leftovers in ice pop molds for a refreshing frozen treat. SERVES 2 (1½ CUPS EACH)

directions

Combine all of the ingredients in a blender and blend until smooth.

Divide between 2 glasses and serve, or store in the refrigerator for up to 3 or 4 days.

nutrition facts (per serving)
130 calories, 2 g fat, 1 g saturated fat, 0 g trans fat, 0 mg cholesterol, 75 mg sodium, 23 g carbohydrates, 2 g fiber, 18 g sugar, 7 g protein, 4% vitamin A, 110% vitamin C, 20% calcium, 2% iron

mango coconut
KEFIR

1 cup Pineapple Chia Colada Kefir (page 103)

1 cup frozen cubed mango

1 tablespoon shredded coconut

A blend of tropical flavors gets an added boost of flavor from shredded coconut—which has slowly been moving from "bad guy" food to health food. This is due to new discoveries and a deeper understanding that all saturated fats are not created equally. The main saturated fat in coconut oil is medium-chain triglycerides, which can increase the levels of healthy cholesterol *and* bad (LDL) cholesterol, yet it seems not to negatively impact the ratio of the two. As science continues to sort out the specific details of coconut oil and even as it moves to more of a health food status, keep in mind that just because some is good, more is not necessarily better—but you can still enjoy the coconut flavors of this smoothie every now and then. **SERVES 1 (1½ CUPS)**

directions

Combine all of the ingredients in a blender and blend until smooth.
   Serve, or store in the refrigerator for up to 3 or 4 days.

nutrition facts (per serving)
250 calories, 8 g fat, 5 g saturated fat, 0 g trans fat, 0 mg cholesterol, 50 mg sodium, 44 g carbohydrates, 6 g fiber, 36 g sugar, 5 g protein, 30% vitamin A, 120% vitamin C, 15% calcium, 4% iron

icy ginger
BLUE

1 cup frozen blueberries	1 cup plain seltzer
½ cup Ginger Beer (page 29) or store-bought ginger beer	

The anti-inflammatory impact of ginger is well studied and numerous studies have continued to confirm the effect. This drink is an icy and refreshing way to enjoy ginger beer with an immune boost from the blueberries. **SERVES 1 (1½ CUPS)**

directions

Combine the blueberries and Ginger Beer in a blender and blend until smooth.

   Pour the blueberry-ginger mixture into a glass, finish with the seltzer, and serve.

nutrition facts (per serving)
130 calories, 0 g fat, 0 g saturated fat, 0 g trans fat, 0 mg cholesterol, 0 mg sodium, 33 g carbohydrates, 3 g fiber, 27 g sugar, 1 g protein, 2% vitamin A, 25% vitamin C, 0% calcium, 2% iron

ENERGIZING

The beverage recipes in this section are refreshing and energizing! Some of the recipes blend in caffeine from black tea, coffee, kombucha, green tea, or espresso beans, while others gain their energizing nature from fresh ingredients — like the Lemon Freeze and the Miso Sunrise. Neither recipe includes caffeine yet the refreshing flavor provides an energy boost all of its own.

If you are sensitive to caffeine, opt for decaffeinated coffee or tea for these recipes; they still have some caffeine but much less compared to their full-strength counterparts. Additionally, it is not recommended to serve children any drinks that include caffeine — the American Academy of Pediatrics recommends that caffeine should not be consumed by children or adolescents due to a number of harmful health effects on developing neurologic and cardiovascular systems.

berry
KOMBUCHA

1 cup frozen mixed berries

½ cup Bill's Strawberry Kefir (page 174)

1 cup Cranberry Ginger Kombucha (page 114) or GT's Enlightened Organic Raw Synergy Cosmic Cranberry Kombucha

Creamy kefir and the sweet flavor of mixed berries balance out the tartness from cranberry ginger kombucha for an energizing smoothie. **SERVES 2 (1 CUP EACH)**

directions

Combine the frozen berries and kefir in a blender and blend until smooth. Add water to thin and ease blending, if necessary.

Stir the berry-kefir mixture into the kombucha.

Divide among 2 glasses and serve, or store in the refrigerator for up to 3 or 4 days.

nutrition facts (per serving)
100 calories, 1 g fat, 0 g saturated fat, 0 g trans fat, <5 mg cholesterol, 30 mg sodium, 20 g carbohydrates, 2 g fiber, 13 g sugar, 3 g protein, 4% vitamin A, 40% vitamin C, 10% calcium, 2% iron

blackberry
GREEN TEA &
LEMONADE

1 cup Water Kefir (page 37)

½ cup chilled brewed green tea

3 tablespoons fresh lemon juice (from 1 lemon)

2 teaspoons granulated sugar

½ cup fresh blackberries

8 to 10 ice cubes

2 fresh lemon slices

This refreshing beverage includes caffeine from green tea plus probiotics from kefir. Studies have found that EGCG (epigallocatechin gallate), a type of catechin found in green tea, may help boost metabolism and help keep weight off once you have lost it. Plus the brewed green tea will provide an extra energizing boost from the caffeine. SERVES 2 (1 CUP EACH)

directions

Stir together the kefir, green tea, lemon juice, and sugar.

Divide the blackberries between 2 glasses. Add ice and half of the green tea and lemonade mixture to each and garnish with the fresh lemon slices.

Serve, or store in the refrigerator for up to 3 or 4 days.

nutrition facts* (per serving)

60 calories, 0 g fat, 0 g saturated fat, 0 g trans fat, 0 mg cholesterol, 10 mg sodium, 16 g carbohydrates, 3 g fiber, 11 g sugar, 1 g protein, 6% vitamin A, 4% calcium, 40% vitamin C, 6% iron

* It is difficult to calculate the nutrition facts for water kefir because the nutrition varies based on how much sugar is fermented by the kefir grains during the process.

blackberry
KOMBUCHA

1 cup fresh blackberries

½ cup 100 percent apple juice

½ cup Original Kombucha (page 23) or GT's Enlightened Organic Raw Original Kombucha

4 or 5 ice cubes (optional)

This icy kombucha beverage will give you an energy boost from the caffeine in the black tea plus an antioxidant boost from the blackberries. SERVES 1 (1½ CUPS)

directions

Combine the blackberries and apple juice in a blender and pulse to gently blend the berries.

Stir the berry mixture into the kombucha and add ice if desired.

Serve, or store in the refrigerator for up to 3 or 4 days.

nutrition facts (per serving)

150 calories, ½ g fat, 0 g saturated fat, 0 g trans fat, 0 mg cholesterol, 15 mg sodium, 35 g carbohydrates, 8 g fiber, 23 g sugar, 2 g protein, 6% vitamin A, 60% vitamin C, 6% calcium, 8% iron

mango green
TEA

1 cup frozen cubed mango

1 cup chilled brewed green tea

½ cup low-fat or nonfat plain yogurt or Homemade Yogurt (page 44)

1 scoop (about 2 tablespoons) vanilla protein powder

1 to 2 teaspoons honey (optional)

The fresh flavors of mango and yogurt combine with a green tea base in this protein-rich smoothie. When choosing a protein powder, look for a high-quality protein like whey protein isolate or pea protein isolate. SERVES 2 (1 CUP EACH)

directions

Combine all of the ingredients in a blender and blend until smooth.

Divide among 2 glasses and serve, or store in the refrigerator for up to 3 or 4 days.

nutrition facts (per serving)

130 calories, 1½ g fat, ½ g saturated fat, 0 g trans fat, <5 mg cholesterol, 110 mg sodium, 23 g carbohydrates, 2 g fiber, 20 g sugar, 8 g protein, 15% vitamin A, 10% calcium, 40% vitamin C, 0% iron

creamy
BLUEBERRY POMEGRANATE SMOOTHIE

½ cup fresh blueberries

½ cup 100 percent blueberry pomegranate juice (such as from POM Wonderful)

4 ice cubes

½ cup Apple Ginger Kefir (page 106)

This smoothie combines nutrient-rich blueberries and pomegranate juice and gets a probiotic boost from apple ginger kefir. When buying pomegranate juice, choose 100 percent juice to ensure maximum nutrition benefits. **SERVES 1 (1½ CUPS)**

directions

Combine the blueberries, pomegranate juice, and ice cubes in a blender and blend until smooth.

In a glass, add the kefir and then pour in the icy pomegranate blueberry mixture. Stir to combine.

Serve, or store in the refrigerator for up to 3 or 4 days.

nutrition facts (per serving)

190 calories, 1½ g fat, 0 g saturated fat, 0 g trans fat, 0 mg cholesterol, 85 mg sodium, 43 g carbohydrates, 3 g fiber, 36 g sugar, 4 g protein, 20% vitamin A, 20% calcium, 20% vitamin C, 4% iron

creamy
ICED CHAI

½ cup chai tea concentrate (such as from Oregon Chai)

½ cup plain kefir, Traditional Plain Kefir (page 31), or filmjölk

3 or 4 ice cubes

This is a very simple energizing and healing drink recipe! Chai tea has warm and rich flavors that are rooted in the Ayurvedic tradition from India and feature healing spices like ginger, cinnamon, black pepper, and cardamom. Ayurveda ("life science") is a traditional system of medicine that includes the practice of yoga and the healing power of herbs and spices. The chai tea concentrate has 30 to 35 grams of caffeine per cup, from the black tea base. SERVES 1 (1 CUP)

directions

Whisk together the chai tea concentrate and kefir in a glass, fill with ice, and stir. Serve immediately.

nutrition facts (per serving)
150 calories, 1 g fat, 1 g saturated fat, 0 g trans fat, 0 mg cholesterol, 95 mg sodium, 28 g carbohydrates, 0 g fiber, 25 g sugar, 6 g protein, 6% vitamin A, 15% calcium, 2% vitamin C, 0% iron

double cherry
KOMBUCHA

1 cup frozen dark sweet cherries

½ cup tart cherry juice

½ cup Original Kombucha (page 23) or GT's Enlightened Organic Raw Original Kombucha

1 teaspoon agave nectar (optional)

Sweet cherries add a delightful dark purple color and sweet flavor to this drink. When combined with tart cherry juice, they are an absolutely perfect pairing with kombucha. The agave nectar is optional because you may find that the drink is sweet enough with just the cherries. SERVES 1 (1½ CUPS)

directions

Combine the frozen cherries and tart cherry juice in a blender and blend until smooth.

Fill a glass with the kombucha. Then stir the icy cherry mixture into the kombucha. Taste and sweeten with agave nectar if necessary. Serve, or store in the refrigerator for up to 3 or 4 days.

nutrition facts (per serving)

180 calories, 0 g fat, ½ g saturated fat, 0 g trans fat, 0 mg cholesterol, 20 mg sodium, 42 g carbohydrates, 3 g fiber, 32 g sugar, 2 g protein, 2% vitamin A, 2% calcium, 15% vitamin C, 6% iron

chocolate
ESPRESSO
BEAN KEFIR

½ cup chilled brewed coffee

½ cup plain kefir or Traditional Plain Kefir (page 31)

½ cup light almond milk or Homemade Almond Milk (page 49)

1½ tablespoons homemade dark chocolate sauce (see Note)

9 chocolate-covered espresso beans or 1 shot of espresso

5 or 6 ice cubes

Each tiny espresso bean has about 5 milligrams of caffeine, so adding 9 chocolate-covered beans provides about 45 milligrams of caffeine to the recipe. SERVES 2 (1 CUP EACH)

directions
Combine all of the ingredients in a blender and blend until frothy.

Divide among 2 glasses and serve, or store in the refrigerator for up to 3 or 4 days.

Note: To make homemade dark chocolate sauce: In a small saucepan combine ¼ cup packed brown sugar and 2 tablespoons water. Bring to a boil, stirring constantly. Add ¼ cup unsweetened cocoa powder and stir until smooth. Remove from the heat and add 1 tablespoon unsalted butter and ¼ teaspoon vanilla extract. Store in an airtight container for up to 3 days; reheat before using.

nutrition facts (per serving)
90 calories, 3½ g fat, 2 g saturated fat, 0 g trans fat, <5 mg cholesterol, 75 mg sodium, 13 g carbohydrates, <1 g fiber, 11 g sugar, 4 g protein, 6% vitamin A, 20% calcium, 0% vitamin C, 4% iron

ginger
FREEZE

8 ounces club soda

4 to 6 Ginger Beer (page 29) ice cubes

After making a batch of ginger beer, freeze some in an ice cube tray to make ginger beer ice cubes. Add to club soda for a completely refreshing drink and fun way to enjoy ginger beer! **SERVES 1 (ABOUT 1 CUP)**

directions

Combine the club soda and Ginger Beer cubes in a glass.
Serve immediately.

nutrition facts (per serving)
50 calories, 0 g fat, 0 g saturated fat, 0 g trans fat, 0 mg cholesterol, 0 mg sodium, 12 g carbohydrates, 0 g fiber, 12 g sugar, 0 g protein, 0% vitamin A, 0% calcium, 0% vitamin C, 0% iron

hemp
SHAKE

¾ cup plain cultured almond milk
yogurt or Nondairy Yogurt (page 47)

1 banana

½ cup light almond milk or
Homemade Almond Milk (page 49)

1 tablespoon hemp powder

4 or 5 ice cubes

2 teaspoons honey (optional)

Hemp powder has an impressive nutrition profile, providing minerals like magnesium, potassium, and iron. Each serving of this shake has 13 percent of your daily requirement of magnesium, 6 percent of iron, and 264 milligrams of potassium. The taste profile of hemp powder is quite strong—balancing it out with fresh fruit works well.
SERVES 2 (1 CUP EACH)

directions

Combine the yogurt, banana, almond milk, and hemp powder in a blender and blend until smooth. Add the ice cubes and honey (if desired) and blend until frothy.

Divide among 2 glasses and serve, or store in the refrigerator for up to 3 or 4 days.

nutrition facts (per serving)
150 calories, 2½ g fat, 0 g saturated fat, 0 g trans fat, 0 mg cholesterol, 70 mg sodium, 31 g carbohydrates, 4 g fiber, 15 g sugar, 3 g protein, 4% vitamin A, 20% calcium, 8% vitamin C, 6% iron

probiotic
LEMONADE

1 quart Water Kefir (page 37)
¼ cup fresh lemon juice

1 lemon, sliced and seeded

By itself, water kefir has a light lemon flavor and a hint of sweetness, making it the perfect starter for a batch of homemade lemonade! The best part about this lemonade is that it gains all its sweetness from the water kefir and needs no added sugar. SERVES 4 (ABOUT 1 CUP EACH)

directions
Combine the kefir, lemon juice, and lemon slices in a container.
  Serve over ice, or store in the refrigerator for up to 3 or 4 days.

nutrition facts* (per serving)
50 calories, 0 g fat, 0 g saturated fat, 0 g trans fat, 0 mg cholesterol, 10 mg sodium, 15 g carbohydrates, 2 g fiber, 9 g sugar, <1 g protein, 10% vitamin A, 4% calcium, 40% vitamin C, 10% iron

* It is difficult to calculate the nutrition facts for water kefir because the nutrition varies based on how much sugar is fermented by the kefir grains during the process.

miso
SUNRISE

2 to 3 kale leaves, chopped

1 carrot, chopped; or ¼ cup 100 percent carrot juice

2 cups cold water

2 teaspoons organic mellow white miso

2 teaspoons minced fresh ginger

1 teaspoon tamari

¼ teaspoon garlic powder

Miso takes center stage here, delivering a somewhat savory taste and probiotic boost. This delicious drink also has antioxidants from the garlic powder, carrot, kale, and ginger. SERVES 2 (1½ CUPS EACH)

directions

Combine all of the ingredients in a blender and blend until smooth.

Divide among 2 glasses and serve, or store in the refrigerator for up to 3 or 4 days.

nutrition facts (per serving)

50 calories, 0 g fat, 0 g saturated fat, 0 g trans fat, 0 mg cholesterol, 590 mg sodium, 10 g carbohydrates, 2 g fiber, 2 g sugar, 3 g protein, 220% vitamin A, 6% calcium, 70% vitamin C, 4% iron

lemon FREEZE

1 cup Probiotic Lemonade (page 96)   2 lemon slices
20 ice cubes (about 2 cups)

A whipped lemon freeze can help energize the afternoon—plus, it is so simple! Make a double or triple batch and freeze leftovers in ice pop makers, a perfect alternative to traditional sugar-laden frozen treats. **SERVES 2 (1 CUP EACH)**

directions

Combine the lemonade and ice in a blender and blend until smooth.

Divide between 2 glasses, garnish with the lemon, and serve, or store in the refrigerator for up to 3 or 4 days.

nutrition facts (per serving)
50 calories, 0 g fat, 0 g saturated fat, 0 g trans fat, 0 mg cholesterol, 10 mg sodium, 15 g carbohydrates, 2 g fiber, 9 g sugar, <1 g protein, 10% vitamin A, 4% calcium, 40% vitamin C, 10% iron

mocha
KEFIR

¾ cup chilled brewed coffee

½ cup plain kefir or Traditional Plain Kefir (page 31)

1 tablespoon agave nectar

1 tablespoon homemade dark chocolate sauce (see Note on page 93)

1½ teaspoons unsweetened cocoa powder

1 teaspoon vanilla extract

An icy mocha beverage with only 110 calories! One of the key ingredients is your own homemade chocolate sauce, which helps make this drink extra chocolaty. SERVES 2 (1 CUP EACH)

directions

Combine all of the ingredients in a blender and blend until smooth.

Divide among 2 glasses and serve, or store in the refrigerator for up to 3 or 4 days.

nutrition facts (per serving)
110 calories, 2 g fat, 1½ g saturated fat, 0 g trans fat, <5 mg cholesterol, 35 mg sodium, 21 g carbohydrates, 2 g fiber, 19 g sugar, 4 g protein, 4% vitamin A, 8% calcium, 4% vitamin C, 4% iron

pineapple chia
COLADA KEFIR

1 cup cubed fresh pineapple

½ cup light coconut milk

½ cup plain kefir or Traditional Plain Kefir (page 31)

2 teaspoons chia seeds

Pineapples are naturally occurring sources of bromelain, an enzyme that could help with arthritis pain (and is what gives pineapple its ability to break down protein, making it a great meat tenderizer). The pineapple also contributes vitamin C to this kefir; each serving delivers 120 percent of our daily requirement of vitamin C (74 milligrams). SERVES 2 (1 CUP EACH)

directions

Combine all of the ingredients in a blender and blend until smooth.

Divide among 2 glasses and serve, or store in the refrigerator for up to 3 or 4 days.

nutrition facts (per serving)
255 calories, 8 g fat, 5 g saturated fat, 0 g trans fat, 0 mg cholesterol, 50 mg sodium, 44 g carbohydrates, 6 g fiber, 36 g sugar, 5 g protein, 30% vitamin A, 15% calcium, 120% vitamin C, 4% iron

mocha
SHAKE

½ cup vanilla nonfat frozen yogurt (with live and active cultures), such as Stonyfield Gotta Have Vanilla frozen yogurt

½ cup chilled brewed coffee

1½ teaspoons unsweetened cocoa powder

This shake will keep you out of the drive-through window and still satisfy your mocha fix! It's made with a blend of frozen yogurt and coffee, with a little chocolate flavor from unsweetened cocoa powder. For an even more intense chocolate flavor, you could substitute chocolate frozen yogurt for the vanilla. **SERVES 1 (1 CUP)**

directions

Combine the frozen yogurt, coffee, and cocoa powder in a blender and blend until smooth.

Serve, or store in the refrigerator for up to 3 or 4 days.

nutrition facts (per serving)
110 calories, 0 g fat, 0 g saturated fat, 0 g trans fat, <5 mg cholesterol, 70 mg sodium, 21 g carbohydrates, <1 g fiber, 19 g sugar, 5 g protein, 0% vitamin A, 15% calcium, 0% vitamin C, 2% iron

HEALING

Let food be thy medicine and medicine be thy food.
— Hippocrates

When you start to look at the many ways that food can help the body, it is absolutely incredible. While sometimes it is necessary to use traditional medicine, food can, in many ways, be medicine to the body. Kefir is used in many recipes throughout this book, but it takes center stage here in the healing section because of its germ-fighting powers. Even everyday foods like raisins have been linked to healing properties. Plus many plant foods' powerful compounds have healing and protective effects on the body, like papaya, which contains zeaxanthin and lutein.

apple ginger
KEFIR

1 apple, such as Honeycrisp, cut into quarters and cored

½ cup plain kefir or Traditional Plain Kefir (page 31)

½ cup light almond milk or Homemade Almond Milk (page 49)

1 teaspoon chopped fresh ginger

1½ teaspoons organic coconut palm sugar

4 or 5 ice cubes

One of the many healthful properties of kefir is its germ-fighting power. Researchers have observed that kefir has antimicrobial power against harmful organisms including *E. coli*, salmonella, shigella, and other pathogenic bacteria. In addition to kefir, this drink includes fresh ginger, which is linked to fighting pain and inflammation. SERVES 2 (1 CUP EACH)

directions

Combine the apple, kefir, almond milk, and ginger in a blender (see Note) and blend until smooth. Add the sugar and ice cubes and blend until frothy.

Divide between 2 glasses and serve, or store in the refrigerator for up to 3 or 4 days.

Note: This works best with a high-speed blender. If using a regular kitchen blender, cut the apple and ginger into smaller pieces before blending.

nutrition facts (per serving)
140 calories, 2 g fat, 1 g saturated fat, 0 g trans fat, 0 mg cholesterol, 140 mg sodium, 26 g carbohydrates, 3 g fiber, 23 g sugar, 6 g protein, 10% vitamin A, 40% calcium, 10% vitamin C, 2% iron

spiced apple
KOMBUCHA

1 apple, cored and sliced

1 cup Original Kombucha (page 23) or GT's Enlightened Organic Raw Original Kombucha

¼ cup chai tea concentrate (such as from Oregon Chai); or 1 teaspoon chai tea spice

6 ice cubes

Kombucha, the fermented wonder drink that has been around for thousands of years, has present-day anecdotal benefits ranging from detoxification to fighting diseases like cancer and arthritis. One animal study from the 1990s shows the evidence for the possible control of different stages of cancer growth with glutaric acid, which is found in kombucha. For a more smoothie-like texture, opt for the blending option. SERVES 2 (ABOUT 1 CUP EACH)

directions

Fill each of 2 glasses with half of the apple slices and ½ cup kombucha. Let the mixture sit for 20 to 30 minutes to absorb some of the apple flavor.

Remove the apple slices. Stir half of the chai tea concentrate (or spice) into each glass and combine. Finish by adding ice cubes to each.

OR

Combine the apple slices, kombucha, chai tea concentrate (or spice), and ice cubes in a blender and blend to combine.

Divide between 2 glasses and serve, or store in the refrigerator for up to 3 or 4 days.

nutrition facts (per serving)
80 calories, 0 g fat, 0 g saturated fat, 0 g trans fat, 0 mg cholesterol, 20 mg sodium, 20 g carbohydrates, 1 g fiber, 13 g sugar, 0 g protein, 0% vitamin A, 0% calcium, 4% vitamin C, 0% iron

tart cherry
KEFIR

1 cup frozen dark cherries

½ cup plain kefir or Traditional Plain Kefir (page 31)

½ cup 100 percent tart cherry juice

4 or 5 ice cubes

Tart cherries are known for their anti-inflammatory impact on the body. Researchers from the Oregon Health & Science University presented a small study of 20 women with inflammatory arthritis (osteoarthritis); it showed that having 8 ounces of tart cherry juice twice a day for 3 weeks led to a reduction in inflammation markers. Similar inflammation reduction has been found among athletes who added tart cherry juice while training for long distance running, and the runners experienced less pain. When shopping for tart cherry juice, opt for 100 percent tart cherry juice to gain the maximum health benefits. **SERVES 2 (1 CUP EACH)**

directions

Combine all of the ingredients in a blender and blend until icy and smooth.

Divide between 2 glasses and serve, or store in the refrigerator for up to 3 or 4 days.

nutrition facts (per serving)
110 calories, ½ g fat, 0 g saturated fat, 0 g trans fat, 0 mg cholesterol, 40 mg sodium, 22 g carbohydrates, 2 g fiber, 18 g sugar, 4 g protein, 4% vitamin A, 8% calcium, 10% vitamin C, 4% iron

chocolate
CHERRY
SMOOTHIE

3 ounces (about ⅕ block) silken tofu

¼ cup low-fat or nonfat plain Greek yogurt

1 tablespoon unsweetened cocoa powder

1 banana

½ cup tart cherry juice

This smoothie gains its healing properties from antioxidant powerhouses: tofu, cocoa powder, berries, and tart cherry juice. In fact, there is an approved health claim by the Food and Drug Administration for soy protein: 25 grams of soy protein a day, as part of a diet low in saturated fat and cholesterol, may reduce the risk of heart disease. The tofu in this smoothie provides about 4 grams soy protein per serving. SERVES 2 (1 CUP EACH)

directions

Combine the tofu, yogurt, and cocoa powder in a blender and blend until smooth. Add the banana and cherry juice and blend again until smooth.

Divide between 2 glasses and serve, or store in the refrigerator for up to 3 or 4 days.

nutrition facts (per serving)
130 calories, 1½ g fat, 0 g saturated fat, 0 g trans fat, 0 mg cholesterol, 25 mg sodium, 25 g carbohydrates, 2 g fiber, 15 g sugar, 6 g protein, 2% vitamin A, 10% calcium, 6% vitamin C, 6% iron

cinnamon
SWIRL

1 cup low-fat or nonfat plain yogurt or Homemade Yogurt (page 44)

½ cup low-fat or nonfat milk

½ cup raisins

½ cup chilled cooked brown rice

1 teaspoon ground cinnamon

4 or 5 ice cubes

One of the healing properties of this beverage is from the raisins: Research presented at the American College of Cardiology's 61st Annual Scientific Session in 2012 found that when men and women were randomly assigned to snack on raisins compared to commercial snacks that did not contain raisins, those snacking on raisins had a reduction of blood pressure. The potential connections: Raisins are loaded with potassium, plus they are natural sources of polyphenols, phenolic acid, and antioxidants. SERVES 2 (1 CUP EACH)

directions

Combine the yogurt, milk, raisins, and rice in a blender and blend until smooth. Add the cinnamon and ice and blend again, until smooth.

Divide between 2 glasses and serve, or store in the refrigerator for up to 3 or 4 days.

nutrition facts (per serving)
160 calories, 1½ g fat, 0 g saturated fat, 0 g trans fat, 0 mg cholesterol, 90 mg sodium, 26 g carbohydrates, 2 g fiber, 11 g sugar, 11 g protein, 2% vitamin A, 25% calcium, 2% vitamin C, 6% iron

banana papaya
SMOOTHIE

1 banana
1 cup cubed papaya

½ cup Water Kefir (page 37)
4 or 5 ice cubes

Papaya contains two xanthophylls—zeaxanthin and lutein—that are important for healthy eyes. It has been found that these two compounds are particularly beneficial in decreasing the risk for age-related macular degeneration. SERVES 2 (1 CUP EACH)

directions

Combine all of the ingredients in a blender and blend until smooth.

Divide between 2 glasses and serve, or store in the refrigerator for up to 3 or 4 days.

nutrition facts* (per serving)

100 calories, 0 g fat, 0 g saturated fat, 0 g trans fat, 0 mg cholesterol, 5 mg sodium, 24 g carbohydrates, 3 g fiber, 15 g sugar, 1g protein, 15% vitamin A, 2% calcium, 80% vitamin C, 2% iron

* It is difficult to calculate the nutrition facts for water kefir because the nutrition varies based on how much sugar is fermented by the kefir grains during the process.

cranberry
GINGER
KOMBUCHA

½ cup Original Kombucha (page 23) or GT's Enlightened Organic Raw Original Kombucha

½ cup frozen whole cranberries

1 teaspoon chopped fresh ginger

Cranberries may not be as well known for their antioxidant capacity as blueberries, but they have antioxidant power that ranks up there near blueberries. Probiotics from the kombucha and an added healing enhancement from the ginger in this recipe add to the cranberries' healing properties. **SERVES 1 (1 CUP)**

directions
Combine the kombucha, cranberries, and ginger in a blender and blend until icy and smooth. If needed, add 1 tablespoon of water at a time to ease blending.

Serve, or store in the refrigerator for up to 3 or 4 days.

nutrition facts (per serving)
60 calories, 0 g fat, 0 g saturated fat, 0 g trans fat, 0 mg cholesterol, 5 mg sodium, 15 g carbohydrates, 4 g fiber, 5 g sugar, 0 g protein, 2% vitamin A, 0% calcium, 20% vitamin C, 2% iron

creamy
BLACKBERRY SMOOTHIE

1 cup fresh blackberries

½ cup low-fat buttermilk

1 teaspoon honey

This smoothie gets a probiotic boost from buttermilk and healing properties from blackberries. Blackberries are loaded with important vitamins including vitamin C (which is linked to helping wounds heal) and vitamin K (which plays a major role in blood clotting). In fact, vitamin K is applied to bruises, scars, and stretch marks to remove them; and following surgery, vitamin C is used to speed up wound healing. SERVES 1 (1½ CUPS)

directions

Combine all of the ingredients in a blender and blend until smooth (there will still be some seeds from the blackberries).

Serve, or store in the refrigerator for up to 3 or 4 days.

nutrition facts (per serving)
130 calories, 2 g fat, ½ g saturated fat, 0 g trans fat, 0 mg cholesterol, 130 mg sodium, 25 g carbohydrates, 8 g fiber, 19 g sugar, 6 g protein, 6% vitamin A, 20% calcium, 50% vitamin C, 6% iron

go
BANANAS

2 frozen bananas

1 cup low-fat or nonfat milk

½ cup Walnut Honey Kefir (page 130)

This smoothie could also be called Potassium Power: Each serving has 537 milligrams of potassium, which plays important roles in the body including maintaining normal body growth. SERVES 2 (1 CUP EACH)

directions

Combine all of the ingredients in a blender and blend until smooth and icy.

Divide between 2 glasses and serve, or store in the refrigerator for up to 3 or 4 days.

nutrition facts (per serving)
160 calories, 3½ g fat, 0 g saturated fat, 0 g trans fat, 0 mg cholesterol, 120 mg sodium, 31 g carbohydrates, 4 g fiber, 4 g sugar, 17 g protein, 8% vitamin A, 30% calcium, 20% vitamin C, 4% iron

honey almond
SHAKE

1 cup low-fat or nonfat plain yogurt or Homemade Yogurt (page 44)

¼ cup light almond milk or Homemade Almond Milk (page 49)

2 tablespoons almond cream (see Note, opposite)

1 to 2 tablespoons dark honey (see Note below)

8 ice cubes

Our bodies have damaging compounds known as free radicals, which can contribute to the process of aging and disease. Including a variety of antioxidants in your daily routine can allow the body to counteract these free radicals. This shake gains its healing powers from honey, which has been found to have antioxidant properties; particularly, the darker the honey the more antioxidant power it has. Honey has also long been recognized for its healing powers, from traditional Ayurveda medicine that used honey to treat imbalances in the body, to the ancient Greeks who believed that honey could make you live longer. When shopping for honey, opt for darker varieties to maximize the antioxidant benefit. SERVES 2 (1 CUP EACH)

directions

Combine all of the ingredients in a blender and blend until frothy and smooth.

Divide between 2 glasses and serve, or store in the refrigerator for up to 3 or 4 days.

Note: Start with 1 tablespoon of honey and, depending on your taste preference, add up to 1 tablespoon more if a little more sweetness is desired.

Note: How to make almond cream: Place 1 cup raw unsalted almonds in a container or bowl and cover entirely with water. Cover and refrigerate overnight. Drain the almonds, transfer to a food processor, and blend until creamy and smooth, 3 to 5 minutes. During blending, add 1 tablespoon of water at a time to thin out the mixture. Store the almond cream in an airtight container in the refrigerator for up to 7 days.

nutrition facts (per serving)
130 calories, 4 g fat, 1½ g saturated fat, 0 g trans fat, 5 mg cholesterol, 100 mg sodium, 19 g carbohydrates, 0 g fiber, 18 g sugar, 6 g protein, 2% vitamin A, 25% calcium, 2% vitamin C, 2% iron

acai melon
SHAKE

2 cups fresh cubed watermelon

½ cup acai berry juice (such as R.W. Knudsen Family Organic acai berry juice blend)

½ cup low-fat or nonfat plain yogurt or Homemade Yogurt (page 44)

4 or 5 ice cubes

The acai berry, a black-purple round berry native to South America, gained super food status due to its antioxidant potential. It can be difficult to find pure acai juice; most often you will find acai juice blends that are a mixture of acai juice and other fruit juices, and sometime sweeteners. A brand to look for is R.W. Knudsen, which is a blend of juices, including organic acai juice, and is 100 percent fruit juice without any added sweeteners. **SERVES 2 (1 CUP EACH)**

directions

Combine all of the ingredients in a blender and blend until smooth.

Divide between 2 glasses and serve, or store in the refrigerator for up to 3 or 4 days.

nutrition facts (per serving)
100 calories, 0 g fat, 0 g saturated fat, 0 g trans fat, 0 mg cholesterol, 30 mg sodium, 20 g carbohydrates, <1 g fiber, 17 g sugar, 7 g protein, 15% vitamin A, 8% calcium, 20% vitamin C, 2% iron

papaya
KEFIR

1 cup cubed fresh papaya

¾ cup Pineapple Chia Colada Kefir (page 103) or Lifeway Coconut-Chia Kefir

4 or 5 ice cubes

The beautiful yellow-orange color of papaya (also known as *papaw* around the world) provides beta-carotene plus lycopene, a carotenoid. Lycopene has been linked to preventing damage to cells and DNA. The shape of the lycopene molecule is in part what makes it so effective at getting rid of free radicals (which can damage cells and DNA). Drink up and enjoy this refreshing kefir! SERVES 1 (1½ CUPS)

directions

Combine the papaya and kefir in a blender and blend until smooth. Add the ice and blend until frozen and smooth.

Serve, or store in the refrigerator for up to 3 or 4 days.

nutrition facts (per serving)
150 calories, 4½ g fat, 2½ g saturated fat, 0 g trans fat, 0 mg cholesterol, 35 mg sodium, 25 g carbohydrates, 5 g fiber, 17 g sugar, 4 g protein, 35% vitamin A, 10% calcium, 180% vitamin C, 2% iron

pom-pineapple
KOMBUCHA

1 cup frozen pineapple

½ cup 100 percent pomegranate juice (such as from POM Wonderful)

½ cup Original Kombucha (page 23) or GT's Enlightened Organic Raw Original Kombucha

1 teaspoon agave nectar (optional)

When you whip up this drink you won't be able to see the over 100 phytochemicals in the pomegranate juice—but your body will know. Science has linked the phytochemicals in pomegranates to protective roles in the body and a reduction in oxidation. The semi-sweet and tart flavor of the pomegranate plus the sweetness of pineapple makes for a tasty drink. SERVES 2 (ABOUT 1 CUP EACH)

directions

Combine the pineapple and pomegranate juice in a blender and blend until smooth. Add water if necessary to ease blending.

Stir the frozen pom-pineapple mixture with the Original Kombucha and if desired, add agave nectar to sweeten, if desired.

Divide between 2 glasses and serve, or store in the refrigerator for up to 3 or 4 days.

nutrition facts (per serving)
90 calories, 0 g fat, 0 g saturated fat, 0 g trans fat, 0 mg cholesterol, 5 mg sodium, 22 g carbohydrates, 1 g fiber, 17 g sugar, 0 g protein, 0% vitamin A, 2% calcium, 45% vitamin C, 2% iron

pumpkin spice
KEFIR

1 cup Vanilla Bean Smoothie (page 128)

½ cup canned pumpkin puree

½ cup cashew cream (see Note)

1 tablespoon pure maple syrup

¼ teaspoon ground cinnamon

4 or 5 ice cubes

A creamy base of vanilla kefir combined with pumpkin and cashew cream gives this beverage a flavor similar to pumpkin pie filling. Another key is the sweetness coming from the pure maple syrup. Why cashew cream? It is a unique way to add thickness and texture to a smoothie. It is dramatically different from cashew butter, which is similar in texture to peanut butter. Cashew cream is made by soaking cashews overnight to soften, then blending them with some water to whip the cashews into a smooth texture that is surprisingly creamy. **SERVES 2 (1 CUP EACH)**

directions

Combine all of the ingredients in a blender and blend until smooth.

Divide between 2 glasses and serve, or store in the refrigerator for up to 3 or 4 days.

Note: How to make cashew cream: Place 1 cup raw unsalted cashews in a container or bowl and cover entirely with water. Cover and refrigerate overnight. Drain and transfer the cashews to a food processor. Blend until creamy and smooth, 3 to 5 minutes. During blending, add 1 tablespoon of water at a time to thin out the mixture if necessary. Store the cashew cream in an airtight container in the refrigerator for up to 7 days.

nutrition facts (per serving)
210 calories, 9 g fat, 2 g saturated fat, 0 g trans fat, 0 mg cholesterol, 45 mg sodium, 25 g carbohydrates, 3 g fiber, 16 g sugar, 10 g protein, 190% vitamin A, 15% calcium, 6% vitamin C, 15% iron

purple power
SHAKE

1 cup frozen sweet dark cherries

½ cup Traditional Plain Kefir (page 31) or plain filmjölk

½ cup 100 percent grape juice

1 tablespoon acai powder

This shake gains its purple power from the sweet dark cherries, 100 percent grape juice, and acai powder. The acai tree, native to tropical Central and South America, has reddish-purple berries that are known to increase energy and health and have gained popularity as a super food. The berry has monounsaturated and polyunsaturated fats and provides essential amino acids plus resveratrol, polyphenols, and flavonoids. Although there is limited evidence based on studies in humans confirming the health benefits of acai, lab studies have demonstrated anti-cancer and anti-inflammatory activity.

SERVES 2 (1 CUP EACH)

directions

Combine the cherries, kefir, and grape juice in a blender and blend until smooth. Add the acai powder and blend until well combined.

Divide between 2 glasses and serve, or store in the refrigerator for up to 3 or 4 days.

nutrition facts (per serving)

120 calories, 1 g fat, 0 g saturated fat, 0 g trans fat, 0 mg cholesterol, 35 mg sodium, 25 g carbohydrates, 2 g fiber, 22 g sugar, 3 g protein, 2% vitamin A, 10% calcium, 10% vitamin C, 2% iron

roasted apple
SMOOTHIE

1 cup Vanilla Bean Smoothie (page 128)

½ cup low-fat or nonfat milk

½ cup roasted cinnamon apples (see Note)

1 apple with peel, cored and sliced

4 or 5 ice cubes

The secret to this smoothie lies in the oven-roasted apple slices; roasting the apples adds a delicious flavor to the smoothie. Plus the natural flavor of the Vanilla Bean Smoothie provides a fantastic base. SERVES 2 (1 CUP EACH)

directions

Combine the vanilla smoothie, milk, and roasted apples in a blender (see Note) and blend until smooth. Add the sliced apple and ice cubes and blend until icy and smooth.

Divide between 2 glasses and serve, or store in the refrigerator for up to 3 or 4 days.

Note: How to prepare roasted cinnamon apples: Core and slice 1 medium apple (such as McIntosh, Golden Delicious, or other type good for baking). In a small bowl, toss the apple slices with 1 tablespoon agave nectar and 1 teaspoon ground cinnamon. Transfer to a baking sheet and roast at 350°F for 15 to 20 minutes, until the apples are tender and lightly browned.

nutrition facts (per serving)
140 calories, 1½ g fat, 0 g saturated fat, 0 g trans fat, 0 mg cholesterol, 85 mg sodium, 28 g carbohydrates, 4 g fiber, 22 g sugar, 6 g protein, 6% vitamin A, 20% calcium, 10% vitamin C, 2% iron

vanilla bean
SMOOTHIE

1½ vanilla bean pods

1 cup plain Traditional Plain Kefir
(page 31) or filmjölk

1 cup low-fat or nonfat plain Greek
yogurt

2 tablespoons agave nectar

8 ice cubes

Vanilla beans are the super tiny flavorful beans that reside inside the vanilla bean pod. Using vanilla beans for a pure vanilla flavor is the signature of this smoothie. The only drawback is that vanilla beans can be pricey — but they are worth every penny. Look for vanilla beans in the natural food section of the grocery store or buy online (see Resources on page 247). If you prefer, 2 teaspoons vanilla extract could be substituted for the vanilla beans, although you won't get the vanilla bean "look" to the beverage.

The smoothie, although delicious on its own, is also a versatile base for other smoothies in the book: Mango Banana Smoothie on page 68, Roasted Apple Smoothie on page 127, and Oatmeal Cookie Smoothie on page 200. SERVES 3 (1 CUP EACH)

directions

To remove the seeds from a vanilla bean pod, snip the ends of a pod and gently slice the pod lengthwise to open. With a spoon, gently scape the vanilla beans from the pod.

Combine the seeds from the vanilla bean pods, the kefir, and yogurt in a blender and blend to combine. Add the agave nectar and ice cubes. Blend until frothy and smooth.

Divide among 3 glasses and serve, or store in the refrigerator for up to 3 or 4 days.

nutrition facts (per serving)
110 calories, ½ g fat, ½ g saturated fat, 0 g trans fat, 0 mg cholesterol,
75 mg sodium, 14 g carbohydrates, 0 g fiber, 14 g sugar, 12 g
protein, 4% vitamin A, 20% calcium, 2% vitamin C, 0% iron

walnut honey
KEFIR

1 cup plain kefir or Traditional Plain Kefir (page 31)

½ cup walnuts

½ cup low-fat or nonfat milk

2 teaspoons honey

4 or 5 ice cubes

This smoothie is nutrient packed with omega-3 fats from the walnuts (the nut with the highest source of omega-3s). In addition, the creaminess of kefir and the subtle sweetness of honey make it a soothing way to start the day—or just enjoy as a snack. To cut the calories, blend in 1 cup frozen fruit (like raspberries) and increase to 3 servings, which will drop the calories by about 100 calories per drink. **SERVES 2 (1 CUP EACH)**

directions

Combine the kefir and walnuts in a blender and blend until smooth.
Add the milk, honey, and ice cubes. Blend until frothy and smooth.
Divide between 2 glasses and serve, or store in the refrigerator for up to 3 or 4 days.

nutrition facts (per serving)
320 calories, 18 g fat, 2 g saturated fat, 0 g trans fat, <5 mg cholesterol, 90 mg sodium, 29 g carbohydrates, 2 g fiber, 27 g sugar, 14 g protein, 8% vitamin A, 25% calcium, 2% vitamin C, 6% iron

DETOXING

The idea of detox is to cleanse your body, and there are many forms of detox diets that claim to do that—some are all liquid based, some require special products, and some can do more harm than good. The drinks in this group are not intended to be a detox diet, but rather a set of beverages that includes plenty of colorful vegetables and fruits to help detox your body.

Foods that have been found to have detoxifying properties include basil, ginger, beets, pineapple, apples, lemons, and probiotics. Each recipe in this chapter includes, of course, a probiotic ingredient or beverage, plus each works in fruit and/or vegetables.

Enjoy these recipes as part of an overall balanced eating plan that helps boost your system with the benefits of a wide variety of vegetables and fruits combined with probiotic-rich ingredients.

raspberry-
GINGER SLUSH

1 cup frozen raspberries

1 cup Ginger Beer (page 29) or store-bought ginger beer

½ cup water

A blend of raspberries and ginger beer has a fresh and slightly sweet flavor. Ginger has been found to give the metabolism a boost, flush out waste, and help keep your appetite in check, according to a 2012 pilot study from Columbia University. SERVES 2 (1 CUP EACH)

directions

Combine the raspberries, ginger beer, and water in a blender and blend until icy and smooth.

Divide between 2 glasses and serve.

nutrition facts (per serving)
90 calories, 0 g fat, 0 g saturated fat, 0 g trans fat, 0 mg cholesterol, 0 mg sodium, 23 g carbohydrates, 2 g fiber, 20 g sugar, <1 g protein, 0% vitamin A, 0% calcium, 10% vitamin C, 2% iron

cantaloupe
LASSI

1 cup Vanilla Lassi (page 52) or
Vanilla Bean Smoothie (page 128)

1 cup cubed cantaloupe

4 or 5 ice cubes

By itself, lassi, a creamy, traditional Indian drink, provides 2 percent of the daily value for vitamin A. But when vitamin A–rich cantaloupe cubes are whipped in, the vitamin A content jumps up to 5,511 IU, or 110 percent of your daily requirement. SERVES 1 (1½ CUPS)

directions

Combine all of the ingredients in a blender and blend until frothy and smooth.

Serve, or store in the refrigerator for up to 3 or 4 days.

nutrition facts (per serving)

190 calories, 3½ g fat, 2 g saturated fat, 0 g trans fat, 15 mg cholesterol, 160 mg sodium, 33 g carbohydrates, 3 g fiber, 31 g sugar, 9 g protein, 110% vitamin A, 30% calcium, 100% vitamin C, 2% iron

cherry lime
FIZZ

1 cup frozen dark sweet cherries

2 tablespoons fresh lime juice

1 teaspoon agave nectar (optional)

¼ cup Ginger Kombucha (page 155) or GT's Enlightened Organic Raw Gingerade Kombucha

The pairing of sweet cherries and lime juice works well with kombucha. To stretch this drink further without adding calories, add a cup of lime seltzer. **SERVES 1 (1 CUP)**

directions

Combine the frozen cherries and lime juice in a blender and blend until smooth. Add water to ease blending, if needed. If desired, add in the agave nectar.

Add the kombucha to a glass, stir in the blended frozen cherry mixture, and combine.

Serve, or store in the refrigerator for up to 3 or 4 days.

nutrition facts (per serving)
110 calories, 0 g fat, 0 g saturated fat, 0 g trans fat, 0 mg cholesterol, 0 mg sodium, 26 g carbohydrates, 3 g fiber, 19 g sugar, 1 g protein, 0% vitamin A, 2% calcium, 30% vitamin C, 2% iron

cran-lemon
BASIL
SMOOTHIE

1 cup frozen whole cranberries

1 cup Probiotic Lemonade (page 96)

¼ cup 100 percent orange juice

4 fresh basil leaves

1 teaspoon honey (optional)

1 lemon slice, cut in half

Using whole cranberries as the base gives this lemony smoothie a nutrient boost from the antioxidants in the cranberries; plus each serving boasts 3 grams of fiber. The basil is added because of its anti-bacterial properties for an extra detox boost. Because of the tartness of the cranberries, honey is listed as an option to add sweetness: Taste the smoothie first, you may find that honey is not needed. SERVES 2 (1 CUP EACH)

directions

Combine the frozen cranberries, lemonade, orange juice, and basil leaves in a blender and blend until smooth. If desired, sweeten with honey.

Divide between 2 glasses, garnish each with a lemon slice, and serve; or store in the refrigerator for up to 3 or 4 days.

nutrition facts (per serving)
60 calories, 0 g fat, 0 g saturated fat, 0 g trans fat, 0 mg cholesterol, 5 mg sodium, 16 g carbohydrates, 3 g fiber, 9 g sugar, <1 g protein, 6% vitamin A, 2% calcium, 60% vitamin C, 6% iron

cucumber mint
SMOOTHIE

1 cup low-fat or nonfat plain yogurt or Homemade Yogurt (page 44)

1 cup cubed cucumber

2 tablespoons fresh mint

1 tablespoon fresh lemon juice

1 teaspoon garlic powder

This light and refreshing smoothie was inspired by tzatziki, the garlicky Greek dip made with yogurt and cucumber. To boost the protein content, swap out plain regular yogurt for plain Greek yogurt. SERVES 2 (1 CUP EACH)

directions

Combine all the ingredients in a blender and blend until smooth.

Divide between 2 glasses and serve, or store in the refrigerator for up to 3 or 4 days.

nutrition facts (per serving)

80 calories, 1½ g fat, 1 g saturated fat, 0 g trans fat, 5 mg cholesterol, 80 mg sodium, 12 g carbohydrates, <1 g fiber, 9 g sugar, 6 g protein, 2% vitamin A, 20% calcium, 10% vitamin C, 2% iron

creamy tomato
SMOOTHIE

½ cup tomato juice

½ cup chopped seeded unpeeled cucumber

¼ cup plain kefir or Traditional Plain Kefir (page 31)

Add this to your breakfast routine for a vegetable boost. To save time in the morning, make up a large batch and refrigerate; gently shake or stir before serving.

The lycopene from the tomato juice may provide ample benefits to your health: Finnish researchers found that lycopene had a protective effect against stroke. Men in the study who had the highest levels of lycopene had a 55 percent lower chance of having any type of stroke. The likely connection, researchers concluded, was the protective effect from the lycopene attacking free radicals and reducing inflammation and cholesterol levels.

The 2010 Dietary Guidelines for Americans recommend we consume 2,300 milligrams or less of sodium, and if you are 51 years or older or have certain health conditions the recommendation is 1,500 milligrams or less. To lower the sodium content of this recipe, choose low-sodium tomato juice. SERVES 1 (1 CUP)

directions
Combine all of the ingredients in a blender and blend until smooth.

Serve, or store in the refrigerator for up to 3 or 4 days.

variation
Add ½ teaspoon (or more, to taste) hot sauce to the mixture for a Bloody Mary twist.

nutrition facts (per serving)
90 calories, 1 g fat, 1 g saturated fat, 0 g trans fat, 0 mg cholesterol,
400 mg sodium, 13 g carbohydrates, 1 g fiber, 10 g sugar, 7 g
protein, 10% vitamin A, 15% calcium, 60% vitamin C, 2% iron

green goodness
SMOOTHIE

½ cup stemmed kale

½ cup apple cider

½ apple, cored

2 strawberries, hulled

½ cup low-fat or nonfat vanilla bean Greek yogurt

If you have been reluctant to try green-colored beverages, this is a great one to start with! The sweetness of the strawberries, apple, and apple cider add delicious flavor to the kale. For a detox bonus, the apples contain the soluble fiber pectin, which can help you detox by removing food additives and metals from your body. SERVES 1 (1½ CUPS)

directions

Combine the kale, cider, apple, and strawberries in a blender (see Note) and blend until smooth.

In a glass, stir together the green juice mixture and vanilla yogurt until combined.

Serve, or store in the refrigerator for up to 3 or 4 days.

Note: This works best with a high-speed blender. If using a traditional kitchen blender, chop the kale, strawberries, and apple before blending.

nutrition facts (per serving)
250 calories, 2½ g fat, 1½ g saturated fat, 0 g trans fat, 15 mg cholesterol, 75 mg sodium, 48 g carbohydrates, 3 g fiber, 37 g sugar, 10 g protein, 110% vitamin A, 30% calcium, 190% vitamin C, 8% iron

grapefruit
GINGER BEER

½ medium grapefruit, peeled and roughly chopped

1 cup Ginger Beer (page 29) or store-bought ginger beer

2 or 3 ice cubes

This is simple to whip up and is completely refreshing! The semi-sweet and tart flavor of the grapefruit pairs nicely with the ginger beer. The drink provides 70 percent (44 milligrams) of the daily requirement of vitamin C and 25 percent (1,186 IU) of vitamin A! In addition, the grapefruit provides the antioxidants beta-carotene, lutein, zeaxanthin, and lycopene. **SERVES 1 (1½ CUPS)**

directions
Combine the grapefruit and ginger beer in a blender and blend to puree. Pour into a glass, finish with a few ice cubes, and serve immediately.

nutrition facts (per serving)
140 calories, 0 g fat, 0 g saturated fat, 0 g trans fat, 0 mg cholesterol, 0 mg sodium, 35 g carbohydrates, 1 g fiber, 34 g sugar, <1 g protein, 25% vitamin A, 2% calcium, 70% vitamin C, 0% iron

cran-apple
GINGER FREEZE

1 apple, cored

1 cup Ginger Beer (page 29) or store-bought ginger beer

½ cup frozen whole cranberries

Icy, refreshing and filling—each serving of this freeze has only 100 calories but packs in 3 grams of fiber to help fill you up. SERVES 2 (1 CUP EACH)

directions

Combine all of the ingredients in a blender (see Note) and blend until icy and smooth.

Divide between 2 glasses and serve.

Note: This works well with a high-speed blender. If using a traditional kitchen blender, chop the apple into smaller pieces first.

nutrition facts (per serving)
100 calories, 0 g fat, 0 g saturated fat, 0 g trans fat, 0 mg cholesterol, 0 mg sodium, 26 g carbohydrates, 3 g fiber, 21 g sugar, 0 g protein, 2% vitamin A, 0% calcium, 10% vitamin C, 0% iron

lemon-lime
KEFIR WATER

3 fresh lime slices   ½ cup Water Kefir (page 37)
2 fresh lemon slices   4 or 5 ice cubes

The lemon and lime in this refreshing drink provide 12 milligrams of vitamin C (20 percent of what you need every day). Another benefit is that lemons can help decrease fluid retention. When you are choosing the citrus, look for fresh limes and lemons that are brightly colored. This is best when served chilled. SERVES 1 (1½ CUPS)

directions
In a glass, gently muddle the lime and lemon slices to release the juices. Add the kefir and ice.

Serve, or store in the refrigerator for up to 3 or 4 days.

nutrition facts* (per serving)
45 calories, 0 g fat, 0 g saturated fat, 0 g trans fat, 0 mg cholesterol, 10 mg sodium, 12 g carbohydrates, 2 g fiber, 9 g sugar, <1 g protein, 10% vitamin A, 4% calcium, 20% vitamin C, 8% iron

* It is difficult to calculate the nutrition facts for water kefir because the nutrition varies based on how much sugar is fermented by the kefir grains during the process.

miso tomato
SMOOTHIE

1 cup cherry tomatoes

½ cup vegetable broth

½ cup 100 percent carrot juice

1 teaspoon organic mellow white miso

A fresh tomato beverage with a little zip from miso. Each serving delivers 370 percent (18,746 IU) of the vitamin A you need, in addition to beta-carotene (from the carrot juice), lycopene (from the tomatoes), and lutein (from the kale). To lower the sodium, choose a low-sodium vegetable broth. SERVES 1 (1½ CUPS)

directions

Combine all of the ingredients in a blender and blend until smooth. Serve, or store in the refrigerator for up to 3 or 4 days.

nutrition facts (per serving)
80 calories, ½ g fat, 0 g saturated fat, 0 g trans fat, 0 mg cholesterol, 500 mg sodium, 16 g carbohydrates, 3 g fiber, 11 g sugar, 2 g protein, 370% vitamin A, 4% calcium, 35% vitamin C, 4% iron

pear raspberry
SMOOTHIE

1 container (5.3 ounces) low-fat or nonfat pear Greek yogurt, such as from Chobani (see Note)

1 cup frozen raspberries

1 cup low-fat or nonfat milk

1 pear, cut into quarters and cored

Pears and raspberries are among the fruit with the highest fiber content, so each serving of this smoothie has 6 grams fiber, which helps to fill you up. One cup of raspberries provides 8 grams fiber and 1 pear provides about 5½ grams. **SERVES 3 (1 CUP EACH)**

directions

Combine the yogurt, frozen raspberries, and milk in a blender and blend until smooth. Add the pear and blend until smooth.

Divide among 3 glasses and serve, or store in the refrigerator for up to 3 or 4 days.

Note: If pear-flavored yogurt is not available, instead use ¾ cup vanilla Greek yogurt or Vanilla Bean Smoothie (page 128).

nutrition facts (per serving)
130 calories, 1½ g fat, 0 g saturated fat, 0 g trans fat, 5 mg cholesterol, 55 mg sodium, 25 g carbohydrates, 6 g fiber, 18 g sugar, 8 g protein, 4% vitamin A, 15% calcium, 25% vitamin C, 2% iron

peach ginger
KOMBUCHA

1 cup frozen peach slices or sliced peeled fresh peaches

¼ cup water

1 teaspoon honey

½ cup Ginger Kombucha (page 155) or GT's Enlightened Organic Raw Gingerade Kombucha

Peach and ginger combine perfectly to sweeten kombucha, with an extra boost of sweetness from honey. Look for frozen peach slices that are not packed with added sugar or syrups. If peaches are in season (late June through August), then opt for fresh peaches. Choose peaches that are firm and free of blemishes. You can tell when a peach is ripe by gently applying pressure to the skin, if you can slightly push in on the skin, then it is ripe. SERVES 1 (1½ CUPS)

directions

Combine the peaches, water, and honey in a blender and blend until smooth.

Pour the kombucha into a glass, add the frozen peach mixture, and gently stir to combine. Serve immediately.

nutrition facts (per serving)

100 calories, 0 g fat, 0 g saturated fat, 0 g trans fat, 0 mg cholesterol, 5 mg sodium, 25 g carbohydrates, 3 g fiber, 21 g sugar, 2 g protein, 10% vitamin A, 2% calcium, 20% vitamin C, 2% iron

pineapple
CUCUMBER
SMOOTHIE

1 cup low-fat or nonfat vanilla yogurt   ½ cup cubed fresh pineapple
1 cup sliced cucumber

The saying "cool as a cucumber" probably originated from the idea that cucumbers are so refreshing on account of their high water content (up to 90 percent). That refreshing cucumber works well in this smoothie and is an excellent way to add vegetables to your beverages. In addition, the enzyme bromelain from the pineapple can improve digestion and cleanse the body. SERVES 2 (1 CUP EACH)

directions
Combine all of the ingredients in a blender and blend until smooth. If necessary, add water to thin the mixture to ease blending.

Divide between 2 glasses and serve, or store in the refrigerator for up to 3 or 4 days.

nutrition facts (per serving)
150 calories, 2 g fat, 1½ g saturated fat, 0 g trans fat, 15 mg cholesterol, 60 mg sodium, 24 g carbohydrates, <1 g fiber, 18 g sugar, 9 g protein, 10% vitamin A, 25% calcium, 35% vitamin C, 4% iron

purple
KOMBUCHA

½ cup blueberries

½ cup 100 percent pomegranate juice (such as from POM Wonderful)

4 ice cubes

½ cup Ginger Kombucha (page 155) or GT's Enlightened Organic Raw Gingerade Kombucha

The nutrition boost of this beverage comes from the deep purple color that the blueberries and pomegranate provide, which makes it rich in anthocyanins. Pomegranate juice is rich in flavor as well, and not overly sweet. When buying pomegranate juice, choose 100 percent pomegranate juice for the maximum nutrition benefit. SERVES 1 (1½ CUPS)

directions

Combine the blueberries, pomegranate juice, and ice in a blender and blend until icy and smooth.

Add the kombucha to a glass and stir in the icy blueberry pomegranate mixture.

Serve, or store in the refrigerator for up to 3 or 4 days.

nutrition facts (per serving)
120 calories, 0 g fat, 0 g saturated fat, 0 g trans fat, 0 mg cholesterol, 10 mg sodium, 31 g carbohydrates, 62 g fiber, 22 g sugar, <1 g protein, 0% vitamin A, 10% calcium, 0% vitamin C, 2% iron

red
DRAGON

½ cup chilled roasted chopped beets (see Note)

½ cup low-fat or nonfat plain Greek yogurt

½ cup pomegranate juice (such as from POM Wonderful)

1 teaspoon fresh chopped ginger

One of my very favorite vegetables is roasted beets—I like them just as much when they are blended with yogurt, pomegranate juice, and fresh ginger. If you make roasted beets for dinner, make extra for smoothies throughout the week, they blend well and add a beautiful bright purple-red hue. **SERVES 1 (1 CUP)**

directions

Combine the roasted beets and yogurt in a blender and blend until smooth.

Add the pomegranate juice and fresh ginger and blend until smooth.

Serve, or store in the refrigerator for up to 3 or 4 days.

Note: To roast beets, scrub and chop a medium beet into ½-inch cubes. Tightly seal in tin foil. Roast in a 425°F oven for 30 to 45 minutes, until tender. Remove from the oven and cool slightly. If you have extra, you can refrigerate them to use for smoothies throughout the week.

nutrition facts (per serving)
170 calories, 0 g fat, 0 g saturated fat, 0 g trans fat, 0 mg cholesterol, 125 mg sodium, 30 g carbohydrates, 2 g fiber, 25 g sugar, 13 g protein, 0% vitamin A, 15% calcium, 6% vitamin C, 4% iron

half and
HALF

1 cup water

1 tablespoon fresh lemon juice

1 teaspoon sugar

½ cup Original Kombucha (page 23) or GT's Enlightened Organic Raw Original Kombucha

4 or 5 ice cubes

A favorite drink of mine is the Arnold Palmer: half lemonade and half iced tea. My combination of kombucha tea, lemon juice, and a tiny amount of sugar (only ½ teaspoon per serving) is a lightly sweetened, probiotic-rich twist on the traditional. SERVES 2 (1 CUP EACH)

directions

Combine the water, lemon juice, and sugar in a glass and stir until the sugar is dissolved. Add the kombucha and ice and serve.

nutrition facts (per serving)

15 calories, 0 g fat, 0 g saturated fat, 0 g trans fat, 0 mg cholesterol, 0 mg sodium, 4 g carbohydrates, 0 g fiber, 3 g sugar, 0 g protein, 0% vitamin A, 0% calcium, 4% vitamin C, 0% iron

ginger
KOMBUCHA

1 cup Original Kombucha (page 23) or GT's Enlightened Organic Raw Original Kombucha

2 teaspoons minced fresh ginger

6 ice cubes

Around the world, ginger, with its strong flavor and health-helping properties, has played a role in both cuisine and healing. It has been found to have an impact on the body similar to that of NSAIDs (nonsteroidal anti-inflammatory drugs), giving the rhizome its pain-fighting and inflammation-lowering reputation. SERVES 2 (1 CUP EACH)

directions

Combine the kombucha and ginger in a glass and let the mixture steep for 1 to 2 hours (or overnight in the refrigerator for a stronger ginger flavor).

Strain the kombucha and divide between 2 glasses. Add the ice cubes and serve.

nutrition facts (per serving)

35 calories, 0 g fat, 0 g saturated fat, 0 g trans fat, 0 mg cholesterol, 10 mg sodium, 8 g carbohydrates, 0 g fiber, 2 g sugar, 0 g protein, 0% vitamin A, 0% calcium, 0% vitamin C, 0% iron

NUTRIENT-RICH

Nearly every beverage in this book could land itself in the Nutrient-Rich category. When you are using primarily foods like fruits, vegetables, yogurt, kefir, and kombucha as the basis of your beverages, they are *all* rich in nutrients. But the recipes that stand out among the others are the ones included in this section.

berry beet
SMOOTHIE

1 cup plain kefir or Traditional Plain Kefir (page 31)

1 cup sliced strawberries

1 small beet (about 1½ inches in diameter), peeled

1 to 2 teaspoons agave nectar (optional)

The bold red-purple color of this smoothie packs in important plant compounds, while the strawberries and beets give it 150 percent of your daily requirement of vitamin C (90 milligrams). When choosing beets, look for smooth and round ones that have bright green tops.
SERVES 1 (1½ CUPS)

directions

Combine the kefir, strawberries, and beet in a blender (see Note) and blend until smooth. Taste, and if you think it needs added sweetness, stir in the agave nectar.

Serve, or store in the refrigerator for up to 3 or 4 days.

Note: This recipe works best in a high-speed blender. If you are using a traditional kitchen blender, chop or shred the beet first to ease the blending process.

nutrition facts (per serving)
190 calories, 2½ g fat, 1½ g saturated fat, 0 g trans fat, 0 mg cholesterol, 190 mg sodium, 31 g carbohydrates, 5 g fiber, 24 g sugar, 13 g protein, 10% vitamin A, 35% calcium, 150% vitamin C, 8% iron

banana berry
SMOOTHIE

2 ounces tempeh

½ banana

4 to 5 strawberries, hulled

½ cup light almond milk or
Homemade Almond Milk (page 49)

¼ cup 100 percent orange juice

1 teaspoon agave nectar

4 or 5 ice cubes

This nutrient-rich smoothie packs in over 100 percent of the daily value for vitamin C—80 milligrams! Plus, it delivers important B vitamins like vitamin B$_6$ (pyridoxine), folate, and niacin. B vitamins are an important group of water soluble vitamins that the body uses for the maintenance of healthy brain cells, the breakdown of carbohydrates, and the production of neurotransmitters that regulate mood. SERVES 1 (1½ CUPS)

directions

Combine the tempeh, banana, strawberries, almond milk, orange juice, and agave nectar in a blender. Add the ice cubes and blend to combine.

Serve, or store in the refrigerator for up to 3 or 4 days.

nutrition facts (per serving)

260 calories, 3½ g fat, 1 g saturated fat, 0 g trans fat, 5 mg cholesterol, 75 mg sodium, 47 g carbohydrates, 4 g fiber, 28 g sugar, 13 g protein, 6% vitamin A, 20% calcium, 130% vitamin C, 10% iron

berry green
SMOOTHIE

1 cup packed baby spinach, stemmed

4 large fresh strawberries, hulled

½ cup Probiotic Lemonade (page 96)

1 cup frozen strawberries

Here's a great way to get over 2 cups of fruit and vegetables into your day, plus an extra boost from lemonade kefir, which is rich in probiotics. The smoothie offers 310 percent of the daily value for vitamin C (183 milligrams), 15 percent for iron (3 milligrams), plus 8 grams of belly-filling fiber. And there is no added sugar—all the sweetness you need comes right from the strawberries and the subtle sweetness of the probiotic lemonade. SERVES 1 (1½ CUPS)

directions

Combine the spinach, fresh strawberries, and lemonade in a blender and blend until smooth. Add the frozen strawberries and blend until smooth.

Serve, or store in the refrigerator for up to 3 or 4 days.

nutrition facts (per serving)
140 calories, 1 g fat, 0 g saturated fat, 0 g trans fat, 0 mg cholesterol, 15 mg sodium, 36 g carbohydrates, 8 g fiber, 22 g sugar, 3 g protein, 10% vitamin A, 8% calcium, 310% vitamin C, 15% iron

butternut
SQUASH
SMOOTHIE

½ cup sliced leeks

1 cup chilled roasted butternut squash (see Note)

1 cup vegetable broth

1 teaspoon organic mellow white miso

¼ teaspoon garlic powder

⅛ teaspoon sea salt

A brilliant orange smoothie with 310 percent of your daily requirement of vitamin A (15,628 IU), and it also gives a boost of B vitamins.
SERVES 1 (1½ CUPS)

directions

Heat a small skillet over medium heat. Add the sliced leeks to the dry pan and cook, stirring, until tender, 5 to 7 minutes.

Combine the sautéed leeks, squash, broth, miso, and garlic powder in a blender and blend until smooth. If the mixture is too thick, add water to thin to the desired consistency. Finish with the sea salt.

Serve, or store in the refrigerator for up to 3 or 4 days.

Note: How to roast butternut squash: Place 2 cups frozen butternut squash cubes on a baking sheet and roast at 350°F for 20 to 25 minutes, until lightly browned. Let cool, then refrigerate in an airtight container.

nutrition facts (per serving)
120 calories, ½ g fat, 0 g saturated fat, 0 g trans fat, 0 mg cholesterol, 620 mg sodium, 29 g carbohydrates, 4 g fiber, 6 g sugar, 3 g protein, 310% vitamin A, 10% calcium, 60% vitamin C, 10% iron

silky coco-cado
SMOOTHIE

1½ cups unsweetened coconut milk beverage (such as from So Delicious)

1 avocado, pitted and peeled

1 tablespoon unsweetened cocoa powder

1 tablespoon agave nectar

This smoothie is rich in belly-slimming monounsaturated fat; each serving has 17 grams of fat, mostly healthy fat from the avocado. Plus it gets an antioxidant boost (and chocolate flavor) from unsweetened cocoa powder. In fact, cocoa powder rounds out the top-20 list of the highest antioxidant-containing foods. **SERVES 2 (1 CUP EACH)**

directions

Combine ½ cup of the coconut milk with the avocado in a blender and blend until smooth.

Add the remaining 1 cup coconut milk, the cocoa powder, and agave nectar and blend again, until smooth.

Divide between 2 glasses and serve.

nutrition facts (per serving)
200 calories, 17 g fat, 5 g saturated fat, 0 g trans fat, 0 mg cholesterol, 20 mg sodium, 16 g carbohydrates, 8 g fiber, 6 g sugar, 2 g protein, 10% vitamin A, 8% calcium, 15% vitamin C, 8% iron

veggie patch
SMOOTHIE

1 cup light almond milk or Homemade Almond Milk (page 49)

½ cup sweet potato puree

½ cup frozen vanilla yogurt (such as Stonyfield Gotta Have Vanilla frozen yogurt)

½ cup frozen peaches

½ cup frozen cubed mango

½ cup frozen cubed butternut squash

1 tablespoon agave nectar

This smoothie, packed with veggies (and some fruit too), has a creamy orange color that is loaded with beta-carotene and 190 percent (9,563 IU) of the daily value for vitamin A. The creaminess and sweetness of the frozen yogurt works well to flavor this smoothie.
SERVES 2 (1 CUP EACH)

directions

Combine all of the ingredients in a blender and blend until smooth. The mixture will be thick; if desired, thin with water or additional almond milk.

Divide between 2 glasses and serve, or store in the refrigerator for up to 3 or 4 days.

nutrition facts (per serving)
230 calories, 3 g fat, 1 g saturated fat, 0 g trans fat, <5 mg cholesterol, 135 mg sodium, 49 g carbohydrates, 3 g fiber, 31 g sugar, 5 g protein, 190% vitamin A, 30% calcium, 60% vitamin C, 6% iron

lemon-blue
SMOOTHIE

1 cup frozen blueberries
½ cup water
1 tablespoon fresh lemon juice

1 teaspoon sugar
½ cup low-fat or nonfat plain Greek yogurt

The refreshing combination of lemon and blueberries and the beautiful purple color of this smoothie provide important nutrients that are beyond the food label—specifically, anthocyanins. SERVES 1 (1½ CUPS)

directions

Combine the blueberries, water, lemon juice, and sugar in a blender and blend until smooth. Add the yogurt and blend until creamy.

Serve, or store in the refrigerator for up to 3 or 4 days.

nutrition facts (per serving)
170 calories, ½ g fat, 0 g saturated fat, 0 g trans fat, 0 mg cholesterol, 60 mg sodium, 31 g carbohydrates, 4 g fiber, 23 g sugar, 13 g protein, 2% vitamin A, 15% calcium, 30% vitamin C, 2% iron

fruit punch
SMOOTHIE

¾ cup So Delicious strawberry cultured coconut milk

2 kiwis, peeled

1 banana

1 cup frozen sweet dark cherries

½ cup low-fat or nonfat milk

¼ cup 100 percent white grape juice

A blend of fruits makes for a nutritionally packed smoothie and a delicious sweet taste. Each serving has 140 percent of the daily value for vitamin C (84 milligrams), plus provides a slew of B vitamins: riboflavin, thiamine, niacin, pantothenic acid, pyridoxine, vitamin B_{12}, and folate. **SERVES 2 (1 CUP EACH)**

directions

Combine all of the ingredients in a blender and blend until icy and smooth.

Divide between 2 glasses and serve, or store in the refrigerator for up to 3 or 4 days.

nutrition facts (per serving)
230 calories, 4½ g fat, 3½ g saturated fat, 0 g trans fat, 5 mg cholesterol, 115 mg sodium, 44 g carbohydrates, 5 g fiber, 36 g sugar, 6 g protein, 6% vitamin A, 20% calcium, 140% vitamin C, 2% iron

orange banana
SMOOTHIE

3 ounces tempeh

1 banana

2 tablespoons water

1 cup 100 percent orange juice

Including tempeh in this smoothie packs an impressive amount of protein—10 grams per serving! For an added bonus, the prebiotic properties of the banana work together with the probiotics of the tempeh to boost your overall gut health. SERVES 2 (1 CUP EACH)

directions

Combine the tempeh, banana, and water in a blender and blend until smooth. Add the orange juice and blend until combined.

Divide between 2 glasses and serve, or store in the refrigerator for up to 3 or 4 days.

nutrition facts (per serving)
180 calories, 3½ g fat, ½ g saturated fat, 0 g trans fat, 0 mg cholesterol, 15 mg sodium, 31 g carbohydrates, 3 g fiber, 18 g sugar, 10 g protein, 2% vitamin A, 4% calcium, 140% vitamin C, 8% iron

pistachio
KEFIR

1 cup plain kefir or Traditional Plain
Kefir (page 31)

½ cup shelled unsalted pistachios

1 frozen banana

½ cup low-fat or nonfat milk

This drink is so filling—it is an excellent nutrient-rich snack or drink. Each serving has 15 grams of mostly healthy fat, coming mainly from the pistachios, and 5 grams of fiber. SERVES 2 (¾ CUP EACH)

directions

Combine the kefir and pistachios in a blender and blend until smooth. Add the banana and milk and blend until smooth.

Divide between 2 glasses and serve, or store in the refrigerator for up to 3 or 4 days.

nutrition facts (per serving)
300 calories, 15 g fat, 3 g saturated fat, 0 g trans fat, <5 mg cholesterol, 90 mg sodium, 31 g carbohydrates, 5 g fiber, 19 g sugar, 15 g protein, 10% vitamin A, 25% calcium, 15% vitamin C, 8% iron

strawberry
SPINACH
POWER

1½ cups Walnut Honey Kefir (page 130), or plain or vanilla low-fat or nonfat yogurt

5 strawberries

½ cup packed baby spinach

4 or 5 ice cubes

Popeye certainly ate his spinach, but generally, most people could squeeze more dark leafy greens into their eating routines. Since spinach can go bad quickly, either make up another batch of this smoothie and have it the next day or sauté the leftover spinach with garlic and a light drizzle of olive oil for 2 to 3 minutes, until the spinach is tender, for a super-quick side dish to go with dinner. Each serving of this smoothie provides 20 percent of your daily requirements of vitamin A (987 IU), vitamin C (14 milligrams), and calcium (213 milligrams). **SERVES 2 (1 CUP EACH)**

directions

Combine the kefir, strawberries, and spinach in a blender and blend until smooth. Add the ice cubes and blend until icy.

Divide between 2 glasses and serve, or store in the refrigerator for up to 3 or 4 days.

---

nutrition facts (per serving)
110 calories, 4½ g fat, 0 g saturated fat, 0 g trans fat, 0 mg cholesterol, 85 mg sodium, 12 g carbohydrates, 1 g fiber, 10 g sugar, 5 g protein, 20% vitamin A, 20% calcium, 20% vitamin C, 2% iron

super berry
SMOOTHIE

3 ounces silken tofu

½ cup frozen raspberries

½ cup frozen blueberries

½ cup 100 percent tart cherry juice

¼ cup water

¼ cup low-fat or nonfat plain Greek yogurt

1 teaspoon honey (optional)

Each serving of this antioxidant-rich smoothie has about 4 grams of soy protein from the silken tofu, which is linked to helping heart health. SERVES 2 (1 CUP EACH)

directions

Combine the tofu, berries, cherry juice, and water in a blender and blend until icy and smooth. Add the yogurt and blend. If desired, sweeten with the honey.

Divide between 2 glasses and serve, or store in the refrigerator for up to 3 or 4 days.

nutrition facts (per serving)
110 calories, 1½ g fat, 0 g saturated fat, 0 g trans fat, 0 mg cholesterol, 25 mg sodium, 19 g carbohydrates, 3 g fiber, 13 g sugar, 6 g protein, 2% vitamin A, 8% calcium, 20% vitamin C, 4% iron

bill's
STRAWBERRY
KEFIR

1½ cups strawberries, hulled

1 cup plain kefir or Traditional Plain Kefir (page 31)

½ cup light almond milk or Homemade Almond Milk (page 49)

Flavored kefir is a scrumptious way to get your daily dose of probiotics. This is another fresh fruit kefir that is absolutely delicious; in fact, my husband (Bill) picked it to be his very favorite recipe of the book. The light strawberry flavor makes it an excellent base for other drinks, like the Very Berry Cheesecake Smoothie (on page 222) and Berry Kombucha (on page 84). SERVES 2 (1 CUP EACH)

directions

Combine all of the ingredients in a blender and blend until frothy and smooth.

Divide between 2 glasses and serve, or store in the refrigerator for up to 3 or 4 days.

nutrition facts (per serving)
230 calories, 2½ g fat, 1 g saturated fat, 0 g trans fat, 0 mg cholesterol, 110 mg sodium, 48 g carbohydrates, 6 g fiber, 41 g sugar, 7 g protein, 40% vitamin A, 180% vitamin C, 30% calcium, 4% iron

sweet potato
SMOOTHIE

1 cup sweet potato puree

½ cup low-fat or nonfat plain yogurt or Homemade Yogurt (page 44)

¼ cup apple cider

1 scoop (about 2 tablespoons) vanilla protein powder

4 or 5 ice cubes

This smoothie gains its nutrient boost from sweet potatoes—each serving has 200 percent of the daily value for vitamin A (10,178 IU), along with 6,107 micrograms of beta-carotene. SERVES 2 (2 CUPS)

directions

Combine all of the ingredients in a blender and blend until smooth. Divide between 2 glasses and serve immediately.

nutrition facts (per serving)
200 calories, 1½ g fat, ½ g saturated fat, 0 g trans fat, <5 mg cholesterol, 170 mg sodium, 37 g carbohydrates, 3 g fiber, 14 g sugar, 10 g protein, 200% vitamin A, 15% calcium, 70% vitamin C, 6% iron

coconut
STRAWBERRY
WHIP

¾ cup So Delicious strawberry
cultured coconut milk

½ cup light coconut milk

½ cup fresh raspberries

This smoothie has an incredibly smooth texture from coconut milk yogurt (also called cultured coconut milk) combined with light coconut milk. The fat and saturated fat all come from the coconut milk, but remember that the type of fat in coconut milk is a medium-chained triglyceride that the body can immediately use for energy. However, it is not the green light to add coconut milk to everything! Enjoy this smoothie occasionally and balance out the intake of saturated fat throughout the day. **SERVES 1 (1 CUP)**

directions

Combine all of the ingredients in a blender and blend until smooth. Serve, or store in the refrigerator for up to 3 or 4 days.

nutrition facts (per serving)
240 calories, 15 g fat, 12 g saturated fat, 0 g trans fat, 0 mg cholesterol, 30 mg sodium, 27 g carbohydrates, 8 g fiber, 17 g sugar, <1 g protein, 0% vitamin A, 35% calcium, 25% vitamin C, 10% iron

very veggie
KEFIR

1 stalk rainbow chard, chopped (about 1½ cups)

1 medium stalk celery, chopped

½ cup chopped cucumber

½ cup plain kefir or Traditional Plain Kefir (page 31)

½ cup light almond milk or Homemade Almond Milk (page 49)

1 avocado, pitted and peeled

½ cup 100 percent white grape juice

½ cup water

⅛ teaspoon sea salt

Blending and drinking your vegetables is an excellent way to boost overall vegetable intake. Note: Avocado, when exposed to air, will brown because of an enzyme called polyphenol oxidase. If there are leftovers, store them in an airtight container; if the top of the smoothie still turns brown, you can simply skim off the oxidized portion (brown color) and enjoy the bright green smoothie that waits underneath it. SERVES 3 (1 CUP EACH)

directions

Combine the chard, celery, cucumber, kefir, and almond milk in a blender and blend until smooth. Add the avocado, grape juice, water, and salt; blend again, until smooth.

Divide among 3 glasses and serve, or store in the refrigerator for up to 3 or 4 days.

nutrition facts (per serving)
180 calories, 10 g fat, 1½ g saturated fat, 0 g trans fat, 0 mg cholesterol, 250 mg sodium, 18 g carbohydrates, 5 g fiber, 13 g sugar, 5 g protein, 30% vitamin A, 25% calcium, 40% vitamin C, 4% iron

avocado
COCONUT
FREEZE

1 frozen banana

1 avocado, pitted and peeled

1 cup plain kefir or Traditional Plain Kefir (page 31)

½ cup light coconut milk

1 teaspoon honey

1 ounce dark chocolate, shaved

Here's an icy and thick smoothie that is hearty enough for a meal replacement and delicious enough to double as dessert. It combines the probiotic power of kefir with the prebiotic power of banana. The avocado adds belly-slimming monounsaturated fat, while the coconut milk contributes saturated fats that are medium-chained triglycerides—which are readily utilized by the body for energy. SERVES 2 (1 CUP EACH)

directions

Combine the banana, avocado, kefir, coconut milk, and honey in a blender and blend until smooth.

Divide between 2 glasses and top each with half of the shaved dark chocolate. Serve immediately.

nutrition facts (per serving)
380 calories, 23 g fat, 8 g saturated fat, 0 g trans fat, 0 mg cholesterol, 80 mg sodium, 39 g carbohydrates, 8 g fiber, 24 g sugar, 9 g protein, 8% vitamin A, 15% calcium, 25% vitamin C, 6% iron

REFUELING

After an intense workout, your body needs to be refueled with food and fluids. A smoothie or beverage is a great way to quickly begin to replenish the nutrients your tired muscles need. It is best to begin the refueling process as soon after a workout that you can tolerate eating and drinking.

Studies show that a blend of protein *and* carbohydrates is best for refueling. Take in mostly protein without carbohydrates, and you miss a key part of the refueling picture.

When it comes to choosing a refueling beverage, take into consideration how long your workout was — for example, if you worked out for only 20 minutes, stick with a lower-calorie refueling option because the amounts of protein and carbohydrates you need are drastically less compared to refueling needs after an intense 2-hour training session.

The majority of the smoothies and beverages in this section purposely include a blend of carbohydrates and protein to optimally refuel your muscles.

almond butter
AND JELLY
SMOOTHIE

1 cup frozen mixed berries

½ cup light almond milk or Homemade Almond Milk (page 49)

½ cup low-fat or nonfat plain Greek yogurt

2 tablespoons almond butter

1 tablespoon agave nectar (optional)

Almond butter is an excellent way to switch up the nut butters that you consume! This PBJ-inspired smoothie packs in 18 grams of protein and 45 grams of carbohydrates to refuel tired muscles. SERVES 1 (1 CUP)

directions

Combine all of the ingredients in a blender and blend until smooth. Serve, or store in the refrigerator for up to 3 or 4 days.

nutrition facts (per serving)

400 calories, 19 g fat, 1½ g saturated fat, 0 g trans fat, 0 mg cholesterol, 135 mg sodium, 45 g carbohydrates, 5 g fiber, 34 g sugar, 18 g protein, 6% vitamin A, 45% calcium, 25% vitamin C, 10% iron

apple pb
SMOOTHIE

¾ cup plain cultured almond milk yogurt or Nondairy Yogurt (page 47)

1 green apple, cut into quarters and cored

1 cup light almond milk or Homemade Almond Milk (page 49)

1 tablespoon agave nectar

1 tablespoon natural creamy peanut butter

4 or 5 ice cubes

Apples and peanut butter go together so well. Combining them in a smoothie with almond milk yogurt allows for a probiotic boost. Since the smoothie has only 150 calories per serving, it's a great refueling option after a lighter workout. SERVES 3 (1 CUP EACH)

directions

Combine the yogurt, apple, and almond milk in a blender (see Note) and blend until smooth. Add the agave nectar and peanut butter and blend until smooth. Last, add the ice cubes and blend until icy and frothy.

Divide among 3 glasses and serve, or store in the refrigerator for up to 3 or 4 days.

Note: This works best with a high-speed blender. If using a regular kitchen blender, chop the apple into small pieces before blending.

nutrition facts (per serving)
150 calories, 4½ g fat, 0 g saturated fat, 0 g trans fat, 0 mg cholesterol, 80 mg sodium, 25 g carbohydrates, 4 g fiber, 16 g sugar, 2 g protein, 4% vitamin A, 20% calcium, 2% vitamin C, 6% iron

apricot protein
BLAST

2 fresh apricots, pitted

½ cup low-fat or nonfat buttermilk

1 scoop (about 2 tablespoons) vanilla protein powder

2 or 3 ice cubes

The tangy flavor of buttermilk is great when blended with the mostly sweet, slightly tart flavor of apricots. Try mixing this smoothie with the Sweet Potato Smoothie on page 175—the apricot and sweet potato flavors are delicious together. SERVES 1 (1 CUP)

directions

Combine all of the ingredients in a blender and blend until smooth. Serve, or store in the refrigerator for up to 3 or 4 days.

nutrition facts (per serving)
150 calories, 2½ g fat, ½ g saturated fat, 0 g trans fat, 0 mg cholesterol, 260 mg sodium, 19 g carbohydrates, 2 g fiber, 14 g sugar, 15 g protein, 30% vitamin A, 15% calcium, 15% vitamin C, 2% iron

banana bread
KEFIR

2 frozen bananas

½ cup plain kefir or Traditional Plain Kefir (page 31)

¼ cup low-fat or nonfat milk, light almond milk, or Homemade Almond Milk (page 49)

½ teaspoon ground cinnamon

The mixture of banana and cinnamon creates a delicious flavor reminiscent of banana bread. This kefir is loaded with 7 grams of fiber and over 900 milligrams (942 milligrams) of potassium! **SERVES 1 (1½ CUPS)**

directions

Combine all of the ingredients in a blender and blend until smooth. Serve, or store in the refrigerator for up to 3 or 4 days.

nutrition facts (per serving)
290 calories, 2½ g fat, 1½ g saturated fat, 0 g trans fat, <5 mg cholesterol, 90 mg sodium, 64 g carbohydrates, 7 g fiber, 38 g sugar, 10 g protein, 10% vitamin A, 25% calcium, 35% vitamin C, 6% iron

banana protein
BLAST

2 ounces tempeh

1 cup light almond milk or Homemade Almond Milk (page 49)

½ banana

1 tablespoon almond butter

1 teaspoon agave nectar

This smoothie will quickly get the job of refueling tired muscles done. A combination of carbohydrates and protein replenishes glycogen stores (the storage form of carbohydrates in the body) after a workout. In fact, replenishing glycogen is a key to having adequate fuel for the next workout! Plus, the blast provides 50 percent of the daily value (500 milligrams) of calcium! SERVES 1 (1½ CUPS)

directions

Combine all of the ingredients in a blender and blend until smooth. Serve, or store in the refrigerator for up to 3 or 4 days.

nutrition facts (per serving)
210 calories, 5 g total fat, 0 g saturated fat, 0 g trans fat, 0 mg cholesterol, 200 mg sodium, 32 g carbohydrates, 4 g fiber, 13 g sugar, 9 g protein, 10% vitamin A, 8% vitamin C, 50% calcium, 10% iron

caramel banana
SMOOTHIE

2 frozen bananas

½ cup low-fat or nonfat plain yogurt or Homemade Yogurt (page 44)

½ cup cashew cream (see Note on page 124)

1 tablespoon caramel sauce

An exercise session lasting longer than 1 hour or in extreme temperatures can deplete the body of potassium. This smoothie helps to replenish your body of the lost potassium that plays crucial roles in the body, including working with sodium to maintain normal blood pressure. SERVES 1 (1½ CUPS)

directions

Combine all of the ingredients in a blender and blend until smooth. Add milk to thin the mixture and ease blending, if needed.

Serve, or store in the refrigerator for up to 3 to 4 days.

nutrition facts (per serving)

210 calories, 5 g total fat, 0 g saturated fat, 0 g trans fat, 0 mg cholesterol, 200 mg sodium, 32 g carbohydrates, 4 g fiber, 13 g sugar, 9 g protein, 10% vitamin A, 8% vitamin C, 50% calcium, 10% iron

chocolate
BANANA
SMOOTHIE
WITH SEA SALT

1 banana

½ cup Vanilla Lassi (page 52) or store-bought vanilla lassi

½ cup low-fat or nonfat milk

1 tablespoon unsweetened cocoa powder

Dash of sea salt (about ⅛ teaspoon, or less)

As you sweat during your workout, run, or game, you lose minerals, including sodium. Unless you have a health issue like high blood pressure and have to limit your sodium intake, as an athlete it is important to include salt in your diet. To finish this refueling drink, you add a dash of sea salt, or about ⅛ teaspoon, which adds about 290 milligrams sodium to the smoothie. If you need to limit your salt intake, omit the sea salt. SERVES 1 (1½ CUPS)

directions

Combine the banana, lassi, milk, and cocoa powder in a blender and blend until the mixture is frothy and smooth. Pour into a glass and then finish with a dash of sea salt.

nutrition facts (per serving)
240 calories, 4 g fat, 2½ g saturated fat, 0 g trans fat, 15 mg cholesterol, 410 mg sodium, 46 g carbohydrates, 6 g fiber, 30 g sugar, 10 g protein, 8% vitamin A, 30% calcium, 20% vitamin C, 6% iron

creamy
BLACKBERRY
GINGER
SMOOTHIE

½ cup low-fat or nonfat plain Greek yogurt

½ cup Original Kombucha, fermented a second time with blackberries (see Second Ferment variation on page 28)

½ teaspoon vanilla extract

⅛ teaspoon ground ginger

Creamy, protein-packed Greek yogurt mixed with blackberry kombucha is an excellent flavor combination. The vanilla extract and ginger add a delicious counterpoint. SERVES 1 (1 CUP)

directions

Stir the yogurt into the kombucha. Add the vanilla and ground ginger.

Mix well and serve immediately.

nutrition facts (per serving)

150 calories, 0 g fat, 0 g saturated fat, 0 g trans fat, 0 mg cholesterol, 60 mg sodium, 23 g carbohydrates, 4 g fiber, 16 g sugar, 13 g protein, 4% vitamin A, 15% calcium, 30% vitamin C, 4% iron

nutella
SMOOTHIE

1 frozen banana

1 cup low-fat or nonfat plain Greek yogurt

¼ cup low-fat or nonfat milk

2 tablespoons Nutella

1 tablespoon agave nectar (optional)

Each serving of this smoothie has a mix of carbohydrates and protein to refuel tired muscles. Plus the banana delivers 257 milligrams of potassium. **SERVES 2 (1 CUP EACH)**

directions

Combine all of the ingredients in a blender and blend until smooth. Serve, or store in the refrigerator for up to 3 or 4 days.

nutrition facts (per serving)
230 calories, 6 g fat, 2½ g saturated fat, 0 g trans fat, 0 mg cholesterol, 75 mg sodium, 30 g carbohydrates, 2 g fiber, 24 g sugar, 15 g protein, 2% vitamin A, 20% calcium, 8% vitamin C, 2% iron

pb cup
SMOOTHIE

1 cup plain kefir or Traditional Plain Kefir (page 31)

¾ cup low-fat or nonfat vanilla yogurt

1½ tablespoons dark chocolate sauce (see Note on page 93)

1½ tablespoons natural peanut butter

4 or 5 ice cubes

What goes together better than chocolate and peanut butter? This smoothie celebrates this iconic flavor combination and blends in probiotics from yogurt and kefir. When choosing peanut butter, it is worth paying extra for natural peanut butter that has an ingredient list of just peanuts and salt. Blended peanut butters have added sugar and oils, giving each serving about 3 grams of sugar — almost a teaspoon worth of added sugar! Once you get used to the more peanut-like flavor of natural peanut butter and stirring it well (the oil separates during storage), you won't turn back. SERVES 2 (1 CUP EACH)

directions

Combine all of the ingredients in a blender and blend until smooth.

Divide between 2 glasses and serve, or store in the refrigerator for up to 3 or 4 days.

nutrition facts (per serving)
280 calories, 12 g fat, 4 g saturated fat, 0 g trans fat, 10 mg cholesterol, 110 mg sodium, 28 g carbohydrates, 2 g fiber, 22 g sugar, 16 g protein, 15% vitamin A, 35% calcium, 10% vitamin C, 6% iron

rice pudding
SMOOTHIE

½ cup chilled cooked brown rice

½ cup low-fat or nonfat vanilla bean yogurt

¼ cup raisins

½ apple, cored

½ cup light almond milk or Homemade Almond Milk (page 49)

¼ cup light coconut milk

1½ teaspoons ground cinnamon

Rice pudding in a glass is a delicious way to enjoy the taste of the iconic comfort dessert. This smoothie gains its sweetness from the raisins, yogurt, and apple—no need for additional sugar. SERVES 2 (1 CUP EACH)

directions

Combine the rice, yogurt, and raisins in a blender (see Note) and blend until smooth. Add the apple, almond milk, coconut milk, and cinnamon. Blend until smooth.

Divide between 2 glasses and serve, or store in the refrigerator for up to 3 or 4 days.

Note: This recipe works best with a high speed blender. If using a traditional blender, chop the apple into smaller pieces before blending.

nutrition facts (per serving)
260 calories, 4 g fat, 2½ g saturated fat, 0 g trans fat, 5 mg cholesterol, 80 mg sodium, 49 g carbohydrates, 4 g fiber, 31 g sugar, 6 g protein, 8% vitamin A, 25% calcium, 10% vitamin C, 6% iron

pb granola
SMOOTHIE

1 banana

½ cup light almond milk or Homemade Almond Milk (page 49)

½ cup low-fat or nonfat plain Greek yogurt

½ cup granola (such as KIND Healthy Grains Oats & Honey Clusters)

1 tablespoon natural peanut butter

1 tablespoon caramel sauce

This smoothie holds up very well when refrigerated; make it ahead for a ready-to-drink post-workout refueling smoothie. The granola and banana create an interesting texture and provide a blend of carbohydrates and protein to refuel tired muscles. SERVES 2 (1 CUP EACH)

directions

Combine all of the ingredients in a blender and blend until smooth (see Note).

Divide between 2 glasses and serve, or store in the refrigerator for up to 3 or 4 days.

Note: Because of the granola, the smoothie will have some texture from the grains.

nutrition facts (per serving)

290 calories, 9 g total fat, 2 g saturated fat, 0 g trans fat, 0 mg cholesterol, 110 mg sodium, 43 g carbohydrates, 5 g fiber, 20 g sugar, 11 g protein, 4% vitamin A, 8% vitamin C, 20% calcium, 8% iron

ants on a log
SMOOTHIE

1 medium stalk celery, chopped, plus optional 2 stalks for garnish

½ cup low-fat or nonfat plain yogurt or Homemade Yogurt (page 44)

1¼ cups low-fat or nonfat milk

2 tablespoons natural peanut butter

¼ cup raisins, plus optional 1 tablespoon for garnish

1 teaspoon honey

The famous kid-friendly snack (celery, PB, and raisins) actually doubles as a delicious-tasting smoothie when mixed with yogurt!
SERVES 2 (1 CUP EACH)

directions

Combine the chopped celery, yogurt, ½ cup of the milk, and the peanut butter in a blender and blend until smooth. Add the remaining ¾ cup milk, the raisins, and honey and blend again until smooth.

Divide between 2 glasses. Serve, garnished with raisins and celery if you like, or store in the refrigerator for up to 3 or 4 days.

nutrition facts (per serving)
220 calories, 10 g total fat, 2 g saturated fat, 0 g trans fat, 0 mg cholesterol, 120 mg sodium, 28 g carbohydrates, 2 g fiber, 24 g sugar, 8 g protein, 6% vitamin A, 2% vitamin C, 30% calcium, 6% iron

raspberry
GRANOLA
SMOOTHIE

1 cup frozen raspberries

1 cup plain coconut milk beverage (such as from So Delicious)

½ cup plain kefir or Traditional Plain Kefir (page 31)

½ cup raspberry granola (such as KIND Healthy Grains Raspberry Clusters with Chia Seeds)

1 teaspoon coconut palm sugar

Adding granola to a smoothie gives it an instant nutrient boost. Make sure you choose granola carefully—not all are created equal. Some are more like a cookie than a health food. Look for varieties that are made with whole grains and have limited added sugar.

SERVES 2 (1 CUP EACH)

directions

Combine the raspberries, coconut milk, and kefir in a blender and blend until smooth. Add the granola and sugar and blend again until smooth (see Note).

Divide between 2 glasses and serve, or store in the refrigerator for up to 3 or 4 days.

Note: Because of the granola, the texture of the smoothie will be a little grainy.

nutrition facts (per serving)
190 calories, 4½ g fat, 2½ g saturated fat, 0 g trans fat, 0 mg cholesterol, 70 mg sodium, 33 g carbohydrates, 6 g fiber, 12 g sugar, 5 g protein, 8% vitamin A, 15% calcium, 30% vitamin C, 8% iron

oatmeal cookie
SMOOTHIE

1 cup Vanilla Bean Smoothie (page 128), store-bought vanilla low-fat or nonfat yogurt, or kefir

¼ cup old-fashioned oats

¼ cup raisins

½ cup low-fat or nonfat milk

¼ teaspoon ground cinnamon

4 or 5 ice cubes

When you are craving a sweet treat, this smoothie is a great option. The sweetness comes from the Vanilla Bean Smoothie and raisins; a hint of cinnamon brings the traditional oatmeal cookie flavor. The 3 grams of fiber and 12 grams of protein will help fill you up and leave you feeling satisfied. SERVES 2 (1 CUP EACH)

directions

Combine the smoothie, oats, and raisins in a blender and blend until smooth (see Note). Add the milk, cinnamon, and ice cubes. Blend again until icy and frothy.

Divide between 2 glasses and serve, or store in the refrigerator for up to 3 or 4 days.

Note: Due to the oats, the texture will be slightly coarse.

nutrition facts (per serving)
220 calories, 2½ g fat, 1 g saturated fat, 0 g trans fat, <5 mg cholesterol, 70 mg sodium, 39 g carbohydrates, 3 g fiber, 25 g sugar, 12 g protein, 4% vitamin A, 20% calcium, 0% vitamin C, 8% iron

strawberry
BANANA
SMOOTHIE

1 frozen banana

1 fresh banana

1 cup Vanilla Bean Smoothie (page 128), store-bought vanilla low-fat or nonfat yogurt, or vanilla kefir

1 cup sliced strawberries

1 to 2 teaspoons agave nectar

A blend of protein from the vanilla smoothie and a nutrient boost from strawberries and bananas make for an excellent refueling drink. Instead of the vanilla bean smoothie, look for store-bought vanilla kefir or yogurt to use as a substitute. **SERVES 2 (1 CUP EACH)**

directions

Combine all of the ingredients in a blender and blend until smooth.

Divide between 2 glasses and serve, or store in the refrigerator for up to 3 or 4 days.

nutrition facts (per serving)
200 calories, 1 g fat, 0 g saturated fat, 0 g trans fat, 0 mg cholesterol, 40 mg sodium, 43 g carbohydrates, 5 g fiber, 28 g sugar, 8 g protein, 4% vitamin A, 10% calcium, 100% vitamin C, 4% iron

blueberry
OMEGA SHAKE

1 cup fresh blueberries

½ cup low-fat or nonfat plain Greek yogurt

½ cup 100 percent blueberry juice

1 tablespoon Barlean's Pomegranate Blueberry Total Omega Swirl flax oil

Blueberries and more blueberries! This shake is a blast of antioxidants and carbohydrates and includes some protein from the Greek yogurt. An additional boost comes from the Barlean's Total Omega Swirl, which is flavored flax oil that provides about 1.6 grams of omega-3 fats. SERVES 1 (ABOUT 1½ CUPS)

directions

Combine the blueberries, yogurt, and juice in a blender and blend until smooth. Add the omega swirl oil and blend until smooth.

Serve, or store in the refrigerator for up to 3 or 4 days.

nutrition facts (per serving)
270 calories, 6 g fat, ½ g saturated fat, 0 g trans fat, 0 mg cholesterol, 60 mg sodium, 45 g carbohydrates, 4 g fiber, 32 g sugar, 13 g protein, 2% vitamin A, 15% calcium, 6% vitamin C, 2% iron

tropical tempeh
SMOOTHIE

2 ounces tempeh

1 cup 100 percent orange juice

½ cup frozen pineapple cubes

2 teaspoons coconut oil

The protein power of tempeh plus the vitamin C from pineapple and orange juice make for a delicious, refreshing, and healing blend that provides 310 percent of the daily value of vitamin C (186 milligrams).
SERVES 1 (1½ CUPS)

directions

Combine all of the ingredients in a blender and blend until smooth. Serve, or store in the refrigerator for up to 3 or 4 days.

nutrition facts (per serving)
320 calories, 12 g fat, 8 g saturated fat, 0 g trans fat, 0 mg cholesterol, 25 mg sodium, 46 mg carbohydrates, 2 g fiber, 28 g sugar, 9 g protein, 4% vitamin A, 310% vitamin C, 10% calcium, 10% iron

THINNING

These recipes are all thinning by the nature of their ingredients, the overall calorie count, or the fiber content. When it comes to losing weight, total calories in (from foods and fluids) and total calories out (from energy expenditure and exercise) still matter.

Stick with the portion size indicated. However, if the recipe says "serves 2" and you have both servings, that may be okay if your overall calorie intake is still in check, just make sure you double the nutrition facts (calories, fat, etc.) for the recipe.

A thinning tip is to have breakfast every day—people who eat breakfast tend to weigh less! If you are prone to breakfast-skipping, try one of these smoothie recipes in a to-go mug for an on-the-go breakfast that will fill you up and get your day started! Or if you have a hectic week, make a batch of smoothies on the weekend and store them in the refrigerator for quick snacks to keep you going all week!

tart cherry–chia
KOMBUCHA

½ cup 100 percent tart cherry juice

½ cup Original Kombucha (page 23) or GT's Enlightened Organic Raw Original Kombucha

1 tablespoon chia seeds

In addition to packing in 5 grams of fiber, this skinny drink boasts chia seeds, which are hydrophilic (meaning they retain water); they actually can absorb ten times their weight in water, which helps you to feel full. Finally, the antioxidant boost that tart cherry juice provides makes this a perfect drink to refuel a tired body. **SERVES 1 (ABOUT 1 CUP)**

directions

Combine the cherry juice, kombucha, and chia seeds in a glass. Stir to combine. Let the beverage set for 5 to 10 minutes (for the chia seeds to start to absorb water).

Serve, or store in the refrigerator for up to 3 or 4 days.

nutrition facts (per serving)
150 calories, 5 g fat, ½ g saturated fat, 0 g trans fat, 0 mg cholesterol, 20 mg sodium, 25 mg carbohydrates, 5 g fiber, 14 g sugar, 3 g protein, 2% vitamin A, 2% vitamin C, 8% calcium, 4% iron

black cherry–
GINGER SODA

1 cup frozen dark sweet cherries

½ cup naturally flavored black cherry seltzer or club soda

½ cup Ginger Beer (page 29) or store-bought ginger beer

Dark sweet cherries and black cherry seltzer balance out the sweetness of ginger beer and add an extra nutrient boost in this refreshing soda. SERVES 1 (1½ CUPS)

directions

Combine the frozen cherries and seltzer in a blender and blend until icy and smooth.

In a glass, add the ginger beer and then stir in the frozen cherry mixture.

Serve immediately.

nutrition facts (per serving)

140 calories, 0 g fat, 0 g saturated fat, 0 g trans fat, 0 mg cholesterol, 0 mg sodium, 34 mg carbohydrates, 3 g fiber, 30 g sugar, 1 g protein, 0% vitamin A, 15% vitamin C, 2% calcium, 2% iron

cherry lime
KOMBUCHA
SODA

½ cup Cherry Lime Fizz (page 134)

¾ cup naturally flavored black cherry seltzer water

This "soda" recipe is refreshing and light in calories with little or no added sugar (depending on whether you included the optional sweetener in the Cherry Lime Fizz). When choosing seltzer water, look for a naturally sweetened variety and avoid those with sugar substitutes like aspartame—the sweetness from the dark sweet cherries and the naturally flavored seltzer is all you will need! SERVES 1 (1½ CUPS)

directions

In a glass, start with the lime fizz, then pour in the seltzer. Gently stir to combine.

Serve immediately.

nutrition facts (per serving)
50 calories, 0 g fat, 0 g saturated fat, 0 g trans fat, 0 mg cholesterol, 0 mg sodium, 13 mg carbohydrates, 2 g fiber, 10 g sugar, <1 g protein, 0% vitamin A, 15% vitamin C, 2% calcium, 2% iron

chia pudding
SMOOTHIE

chia pudding

2 cups light almond milk or Homemade Almond Milk (page 49)

½ cup chia seeds

1 tablespoon pure maple syrup

chia pudding smoothie

½ cup low-fat or nonfat vanilla yogurt

½ teaspoon ground cinnamon

Chia seeds have been popping up on super food lists everywhere, and for a good reason. The tiny seeds are loaded with fiber and are a wonderful plant-based source of omega-3 fats. This smoothie is in the "skinny" smoothie category because of the chia seeds! The fiber from chia helps to fill you up and makes the smoothie a great on-the-go breakfast. Each serving has 14 grams of fiber toward your daily goal of 25 to 30 grams. SERVES 2 (1 CUP EACH)

To make the pudding: Combine the almond milk, chia seeds, and maple syrup in a bowl and stir to combine. Transfer the mixture to a storage container, cover, and refrigerate overnight. The mixture will have a pudding-like consistency. Makes about 2 cups.

To make the smoothie: Combine 1½ cups of the chia pudding,* the yogurt, and cinnamon in a blender and pulse to blend.

Divide between 2 glasses and serve, or store in the refrigerator for up to 3 or 4 days.

* Enjoy the leftover pudding with a dash of cinnamon and chopped apple.

nutrition facts (per serving)

270 calories, 16 g fat, 2 g saturated fat, 0 g trans fat, 0 mg cholesterol, 135 mg sodium, 30 g carbohydrates, 14 g fiber, 14 g sugar, 10 g protein, 10% vitamin A, 6% vitamin C, 60% calcium, 4% iron

creamy
WATERMELON SMOOTHIE

3 cups cubed watermelon

½ cup low-fat or nonfat plain yogurt or Homemade Yogurt (page 44)

8 ice cubes

1 to 2 teaspoons honey (optional)

Blending watermelon with yogurt is a skinny way to enjoy a smoothie with a fresh watermelon flavor. Each serving of this smoothie only has 100 calories and is lightly sweet from the watermelon; add a bit of honey if you find it needs additional sweetness. **SERVES 2 (1½ CUPS EACH)**

directions

Combine the watermelon, yogurt, and ice cubes in a blender and blend until smooth. If additional sweetness is desired, add the honey and blend again until combined.

Divide between 2 glasses and serve immediately.

nutrition facts (per serving)

100 calories, 1 g fat, ½ g saturated fat, 0 g trans fat, <5 mg cholesterol, 40 mg sodium, 21 g carbohydrates, <1 g fiber, 18 g sugar, 4 g protein, 25% vitamin A, 30% vitamin C, 10% calcium, 4% iron

grape
SMOOTHIE

1 cup frozen grapes

1 cup low-fat or nonfat milk

½ cup low-fat or nonfat plain Greek yogurt

2 teaspoons honey (optional)

4 or 5 ice cubes

One cup of grapes has only 55 calories and adds an icy texture and lightly sweet flavor to a smoothie. Frozen grapes are delicious all by themselves as a snack: Wash a bunch of grapes and place in a freezer bag for a staple to keep in your freezer. SERVES 2 (1 CUP EACH)

directions

Combine all of the ingredients in a blender and blend until icy and smooth.

Divide between 2 glasses and serve, or store in the refrigerator for up to 3 or 4 days.

nutrition facts (per serving)

140 calories, 1½ g fat, 1 g saturated fat, 0 g trans fat, 5 mg cholesterol, 80 mg sodium, 23 g carbohydrates, <1 g fiber, 21 g sugar, 11 g protein, 6% vitamin A, 20% calcium, 15% vitamin C, 2% iron

green chia
KEFIR

1 cup plain kefir or Traditional Plain Kefir (page 31)

1 tablespoon chia seeds

1 teaspoon spirulina

1 teaspoon agave nectar

This kefir recipe gets an omega-3 and antioxidant boost with the addition of chia seeds and spirulina. The water-retaining properties of chia seeds, which hold about ten times their weight in water, thicken the kefir. The drink grabs its color from the blue-green algae spirulina, which delivers complete plant-based protein (contains all the essential amino acids) plus iron, vitamin A, and vitamin E.
SERVES 1 (ABOUT 1½ CUPS)

directions
Combine all the ingredients in a glass. Mix to combine and serve.

nutrition facts (per serving)
220 calories, 8 g fat, 2½ g saturated fat, 0 g trans fat, 0 mg cholesterol, 240 mg sodium, 25 g carbohydrates, 6 g fiber, 17 g sugar, 19 g protein, 10% vitamin A, 6% vitamin C, 40% calcium, 20% iron

honeydew lime
SMOOTHIE

1 cup cubed honeydew melon

½ cup low-fat or nonfat plain yogurt or Homemade Yogurt (page 44)

1 tablespoon fresh Key lime juice (from 1 to 2 Key limes; see Note)

4 or 5 ice cubes

This smoothie gains its fresh flavor from Key lime juice, and with only 70 calories. It is a refreshing smoothie that is light in calories and a great addition to a snack or breakfast. Key limes are smaller (about the size of a Ping-Pong ball) and have a notably different flavor compared to traditional limes. They are often available in the grocery store year-round (from Mexico and Central America); but the peak season to find them is June through August. Each lime will yield about ½ to 1 tablespoon of juice. SERVES 2 (1 CUP EACH)

directions

Combine all of the ingredients in a blender and blend until smooth.

Divide between 2 glasses and serve, or store in the refrigerator for up to 3 or 4 days.

Note: Key limes are more perishable than regular Persian limes; store them in the refrigerator in a storage container or bag for up to 5 days.

nutrition facts (per serving)

70 calories, 0 g fat, 0 g saturated fat, 0 g trans fat, 0 mg cholesterol, 40 mg sodium, 11 g carbohydrates, <1 g fiber, 9 g sugar, 6 g protein, 0% vitamin A, 8% calcium, 30% vitamin C, 0% iron

key lime pie
SMOOTHIE

1 cup vanilla frozen yogurt with live and active cultures (such as Stonyfield Gotta Have Vanilla frozen yogurt)

1 tablespoon fresh Key lime juice (from 1 to 2 Key limes)

1½ graham crackers (3 squares)

2 Medjool dates

½ teaspoon vanilla extract

4 or 5 ice cubes

A slice of Key lime pie can have as many as 450 calories; each serving of this smoothie has just about half that number. The distinct flavor of the Key lime juice makes this a perfect dessert-like beverage to enjoy in a glass. Certainly this smoothie could also be made with regular vanilla yogurt instead of the frozen yogurt, yet the frozen yogurt is what really makes the drink! **SERVES 2 (1 CUP EACH)**

directions

Combine the yogurt, lime juice, 2 of the graham cracker squares, and the dates in a blender and blend until smooth. Add the vanilla and ice and blend until icy and frothy.

Crumble the remaining graham cracker square.

Divide the smoothie between 2 glasses and serve, sprinkled with the graham cracker crumbs, or store in the refrigerator for up to 3 or 4 days.

nutrition facts (per serving)
230 calories, 1½ g fat, 0 g saturated fat, 0 g trans fat, <5 mg cholesterol, 120 mg sodium, 48 g carbohydrates, 2 g fiber, 37 g sugar, 5 g protein, 0% vitamin A, 20% calcium, 40% vitamin C, 4% iron

macaroon
SMOOTHIE

1 cup coconut milk beverage (such as from So Delicious)

¼ cup unsweetened cultured coconut milk (such as from So Delicious)

1 tablespoon organic coconut palm sugar

4 or 5 ice cubes

1 tablespoon unsweetened coconut flakes

This smoothie has a light coconut flavor and a filling 110 calories on account of the 8 grams of fat from the coconut and coconut products (milk and yogurt). The interesting thing about the saturated fat in coconut is that it is a medium-chained triglyceride that is directly utilized by the body for energy. Although research is emerging that is identifying the types of saturated fat that need to be limited, remember that while some saturated fat can be part of a balanced eating routine, even saturated fat from coconut products needs to be in moderation. SERVES 2 (1 CUP EACH)

directions

Combine the coconut milk beverage, cultured coconut milk, sugar, and ice in a blender and blend until frothy and smooth.

Heat a small skillet over medium heat. Add the coconut flakes and toast, stirring occasionally, until a light brown color, 3 to 4 minutes.

Divide the smoothie between 2 glasses and serve topped with the toasted coconut, or store in the refrigerator for up to 3 or 4 days.

nutrition facts (per serving)
110 calories, 8 g fat, 7 g saturated fat, 0 g trans fat, 0 mg cholesterol, 10 mg sodium, 12 g carbohydrates, 2 g fiber, 9 g sugar, 0 g protein, 6% vitamin A, 10% calcium, 0% vitamin C, 4% iron

mojito water
KEFIR

4 lime slices, seeded

8 mint leaves

½ cup Water Kefir (page 37)

4 or 5 ice cubes

A mojito traditionally blends together lime slices, mint leaves, simple syrup, and rum—yet it doubles well as a refreshing drink! Add water kefir for a hint of subtle sweetness and skip the rum and you have a refreshing probiotic drink, ideal for a hot day. SERVES 1 (1 CUP)

directions

Muddle the lime slices and mint leaves in a glass to release the juice and mint flavor. Add the kefir and ice. Serve chilled.

nutrition facts* (per serving)

45 calories, 0 g fat, 0 g saturated fat, 0 mg cholesterol, 10 mg sodium, 12 g carbohydrates, 2 g fiber, 9 g sugar, <1 g protein, 10% vitamin A, 4% calcium, 20% vitamin C, 8% iron

* It is difficult to calculate the nutrition facts for water kefir because the nutrition varies based on how much sugar is fermented by the kefir grains during the process.

watermelon
FREEZE

2 cups cubed watermelon

6 ice cubes

1 cup naturally flavored black cherry seltzer

¼ cup Coconut Water Kefir (page 40)

Tomatoes usually steal the stage when it comes to lycopene, but watermelon is rich in lycopene too. In fact, this freeze has 13,777 micrograms of lycopene from the watermelon. The probiotic boost comes from the Coconut Water Kefir. SERVES 1 (ABOUT 1 CUP)

directions

Combine the watermelon and ice in a blender and blend until smooth. Add water during blending if needed to ease blending.

Add the watermelon mixture to a glass, then top with the cherry seltzer and kefir. Stir gently to combine and serve.

nutrition facts* (per serving)

140 calories, ½ g fat, 0 g saturated fat, 0 g trans fat, 0 mg cholesterol, 75 mg sodium, 35 g carbohydrates, 3 g fiber, 27 g sugar, 2 g protein, 45% vitamin A, 6% calcium, 60% vitamin C, 15% iron

* It is difficult to calculate the nutrition facts for water kefir because the nutrition varies based on how much sugar is fermented by the kefir grains during the process.

very berry
CHEESECAKE
SMOOTHIE

1 ounce light cream cheese

1 cup Bill's Strawberry Kefir (page 174)

1 cup frozen mixed berries

1 graham cracker square, crumbled

Who wants cheesecake for breakfast? Of course, this smoothie isn't exactly like traditional cheesecake, but it sure is a great way to enjoy the creamy flavor and texture without as many calories. SERVES 2 (1 CUP EACH)

directions

Combine the cream cheese, kefir, and berries in a blender and blend until smooth.

Divide between 2 glasses, top each with crumbled graham cracker, and serve immediately.

nutrition facts (per serving)
170 calories, 4½ g fat, 2½ g saturated fat, 15 mg cholesterol, 120 mg sodium, 27 g carbohydrates, 3 g fiber, 18 g sugar, 7 g protein, 10% vitamin A, 20% calcium, 70% vitamin C, 6% iron

honey pear
SHAKE

2 pears, cored
½ cup low-fat or nonfat buttermilk

¼ cup water
2 teaspoons honey

Each serving of this shake has 4 grams of fiber and only 140 calories. There are many types of pears to choose from including Bosc, Bartlett, and Asian (which is also known as apple pear), just to name a few. The Bartlett is a great option for this shake because it tends to be very juicy and flavorful, and as it ripens it gains sweetness. Let the pears ripen until they are slightly soft to the touch so that they will blend smoothly. If the pears are too ripe, it can give the shake a grainy texture. SERVES 2 (1 CUP EACH)

directions
Combine all the ingredients in a blender (see Note) and blend until smooth. If the shake is too thick, thin with additional water or ice cubes.

   Divide between 2 glasses and serve, or store in the refrigerator for up to 3 or 4 days.

Note: This works best with a high-speed blender. If using a regular kitchen blender, chop the pears into smaller pieces before blending.

nutrition facts (per serving)
140 calories, 1½ g fat, 1 g saturated fat, 0 g trans fat, <5 mg cholesterol, 55 mg sodium, 31 g carbohydrates, 4 g fiber, 23 g sugar, 3 g protein, 2% vitamin A, 10% calcium, 10% vitamin C, 2% iron

strawberry chia
SMOOTHIE

½ cup Bill's Strawberry Kefir (page 174)

½ cup chilled brewed passion tea (such as from Tazo)

1 tablespoon chia seeds

This smoothie is simple to stir together and is refreshing, light, and yet incredibly filling on account of the 9 grams of fiber from the strawberries and chia. If you prefer an icy texture, blend the ingredients together with 2 or 3 ice cubes. **SERVES 1 (1 CUP)**

directions

Combine the kefir and passion tea in a glass and stir to combine. Add the chia seeds and stir to combine.

Let the mixture set for 10 to 15 minutes for the chia seeds to absorb the liquid. Stir again and serve.

nutrition facts (per serving)

200 calories, 7 g fat, 1 g saturated fat, 0 g trans fat, 0 mg cholesterol, 41 mg sodium, 41 g carbohydrates, 9 g fiber, 30 g sugar, 8 g protein, 30% vitamin A, 30% calcium, 130% vitamin C, 2% iron

COCKTAILS AND SHAKES

It is important to make room in your eating routine for some "fun" calories and treats! These cocktail, soda, and shake recipes are certainly fun, but each also includes a probiotic base like frozen yogurt, frozen kefir, ginger beer, or yogurt, and even one drink includes pickles! Think of these as healthier twists on cocktails and shakes.

When buying store-bought frozen yogurt, make sure to look for the words "live and active cultures" on the packaging or product website, as not all frozen yogurts are created equal. If you have trouble finding frozen yogurt or frozen kefir at your grocery store, as a stand-in freeze homemade yogurt or kefir in ice cube trays and use in the recipes instead of the store-bought.

"jose GRABOWSKI"

supplies
Melon baller

ingredients
1 large dill pickle (such as Bubbies Pure Kosher Dills), cut in half crosswise

2 ounces (¼ cup) tequila

This drink has a fun story behind it—it's a family tradition that my dad and uncles started many, many years ago. The story goes that they were all having a fun night and when it was time to have a shot of tequila they realized they were out of lemons, so they got a little creative and found some pickles; they figured the salty flavor would go well with the tequila. While I'm certain the probiotic power of pickles is not what they had in mind when they created it, this signature family drink has found its way into print! As for the name, Jose is for the Jose Cuervo tequila and Grabowski for the pickle. Enjoying these has become a matter of tradition, one that has been carried out at numerous gatherings and been shared with too many friends to count. SERVES 2 (1 PICKLE SHOT EACH)

directions
Hollow out each half of the pickle using the melon baller, making 2 shot glasses.

Fill each half of the pickle with the tequila and cheers! The pickle is the chaser for the shot: After you have the tequila, eat the pickle.

nutrition facts (per serving)
90 calories, 0 g fat, 0 g saturated fat, 0 g trans fat, 0 mg cholesterol, 870 mg sodium, 3 g carbohydrates, <1 g fiber, 2 g sugar, 0 g protein, 2% vitamin A, 0% calcium, 2% vitamin C, 2% iron

dark and
SKINNY

3 orange slices

1 ounce (2 tablespoons) dark rum (such as Gosling's Black Seal)

4 or 5 ice cubes

4 ounces (½ cup) club soda

2 ounces (¼ cup) Ginger Beer (page 29) or store-bought ginger beer

A Dark and Stormy is a traditional ginger beer–based drink. I like to combine it with some orange slices and add club soda to stretch the serving further and keep the calories to just 130! SERVES 1 (1½ CUPS)

directions

Muddle 2 of the orange slices with the rum in a glass.

Add the ice cubes and top with the club soda and ginger beer.

Garnish with the remaining orange slice and serve.

nutrition facts (per serving)
130 calories, 0 g fat, 0 g saturated fat, 0 g trans fat, 0 mg cholesterol, 0 mg sodium, 13 g carbohydrates, 1 g fiber, 12 g sugar, <1 g protein, 2% vitamin A, 2% calcium, 50% vitamin C, 0% iron

tart cherry gin
AND GINGER

½ cup 100 percent tart cherry juice

½ cup Ginger Beer (page 29) or store-bought ginger beer

1 ounce (2 tablespoons) gin

4 or 5 ice cubes

This is a refreshing cocktail with a nutrient boost from the addition of tart cherry juice. **SERVES 1 (1½ CUPS)**

directions

Combine the cherry juice, ginger beer, and gin in a glass. Mix to combine and add the ice.

Serve immediately.

nutrition facts (per serving)

190 calories, 0 g fat, 0 g saturated fat, 0 g trans fat, 0 mg cholesterol, 15 mg sodium, 29 g carbohydrates, 0 g fiber, 25 g sugar, <1 g protein, 2% vitamin A, 0% calcium, 2% vitamin C, 4% iron

ginger
BELLINI

1 cup frozen peaches (without juice)

1 cup water

1 cup Ginger Beer (page 29) or store-bought ginger beer

2 fresh peach slices for garnish

The Bellini cocktail originated in Venice and traditionally includes sparkling wine and peach puree or nectar. Although you can find 100 percent fruit juice–based nectar, most are more sugar than fruit. Instead, this recipe calls for a homemade puree (frozen peaches blended with water) as the base. SERVES 2 (1½ CUPS EACH)

directions

Combine the frozen peaches and water in the blender and blend until smooth.

Divide the peach puree between 2 highball glasses and stir in the ginger beer. Garnish each glass with a peach slice and serve.

nutrition facts (per serving)
80 calories, 0 g fat, 0 g saturated fat, 0 g trans fat, 0 mg cholesterol, 0 mg sodium, 20 g carbohydrates, 1 g fiber, 19 g sugar, <1g protein, 6% vitamin A, 0% calcium, 10% vitamin C, 2% iron

stormy
GINGER

4 fresh lime slices

1 ounce (2 tablespoons) dark rum
(such as Gosling's Black Seal)

4 ounces (½ cup) Ginger Beer (page
29) or store-bought ginger beer

4 or 5 ice cubes

This drink combines dark rum with a probiotic punch from home-made ginger beer and finishes with a touch of lime. A fun and celebratory way to work in probiotics! **SERVES 1 (1½ CUPS)**

directions

In a highball glass, muddle 3 of the lime slices with the rum.

Fill the glass with the ginger beer and ice. Garnish with the remaining lime slice and serve.

nutrition facts (per serving)

120 calories, 0 g fat, 0 g saturated fat, 0 g trans fat, 0 mg cholesterol,
0 mg sodium, 14 g carbohydrates, 0 g fiber, 13 g sugar, 0 g
protein, 0% vitamin A, 0% calcium, 8% vitamin C, 0% iron

apple ginger
SODA

4 or 5 ice cubes

½ cup apple cider

½ cup Ginger Beer (page 29) or store-bought ginger beer

Combining apple cider with ginger soda is a delicious twist on the traditional ginger soda! To stretch the sweetness further, add 1 cup club soda. SERVES 1 (1½ CUPS)

directions

In a highball glass, add the ice cubes and then top with the cider and ginger beer.

Gently stir to combine all the ingredients and serve immediately.

nutrition facts (per serving)

110 calories, 0 g fat, 0 g saturated fat, 0 g trans fat, 0 mg cholesterol, 0 mg sodium, 26 g carbohydrates, 0 g fiber, 25 g sugar, 0 g protein, 0% vitamin A, 0% calcium, 50% vitamin C, 2% iron

strawberry
GINGER FIZZ

2 large strawberries, sliced

1 ounce (2 tablespoons) gin

4 or 5 ice cubes

2 ounces (¼ cup) club soda

2 ounces (¼ cup) Ginger Beer (page 29) or store-bought ginger beer

This drink is so refreshing: The hint of ginger from the ginger beer tastes delicious with the strawberries, plus the strawberries add 50 percent of your daily requirement of vitamin C. **SERVES 1 (1½ CUPS)**

directions

Muddle all but one of the strawberry slices with the gin in the bottom of a glass.

Add the ice cubes and top with the club soda and ginger beer.

Garnish with the remaining strawberry slice and serve immediately.

nutrition facts (per serving)

120 calories, 0 g fat, 0 g saturated fat, 0 g trans fat, 0 mg cholesterol, 0 mg sodium, 10 g carbohydrates, 1 g fiber, 9 g sugar, 0 g protein, 0% vitamin A, 0% calcium, 50% vitamin C, 2% iron

orange ginger
FIZZ

4 orange slices

1 ounce (2 tablespoons) clementine vodka

4 or 5 ice cubes

4 ounces (½ cup) Ginger Beer (page 29) or store-bought ginger beer

This drink combines the fresh flavors of oranges and clementine vodka with a punch of ginger beer: Such a refreshing cocktail! To "skinny-size" the cocktail, add 4 ounces (½ cup) of club soda and serve in a tall glass. SERVES 1 (1½ CUPS)

directions

Muddle 3 of the orange slices with the vodka in a glass.

Add the ice cubes and top with the ginger beer. Garnish with the remaining orange slice and serve.

nutrition facts (per serving)
150 calories, 0 g fat, 0 g saturated fat, 0 g trans fat, 0 mg cholesterol, 0 mg sodium, 18 g carbohydrates, 1 g fiber, 17 g sugar, 0 g protein, 2% vitamin A, 2% calcium, 40% vitamin C, 0% iron

cupcake
KEFIR

1 cup Vanilla Bean Smoothie (page 128)

1 cup yellow cake cubes (about 2 cupcakes' worth)

5 to 6 strawberries, hulled

4 or 5 ice cubes

If you have leftover birthday cake, this is the perfect drink to whip up! The recipe calls for yellow cake cubes, but chocolate would taste just right too. SERVES 2 (1 CUP EACH)

directions

Combine all of the ingredients in a blender and blend until smooth.

Divide between 2 glasses and serve, or store in the refrigerator for up to 3 or 4 days.

nutrition facts (per serving)
230 calories, 7 g fat, 2 g saturated fat, 0 g trans fat, 25 mg cholesterol, 190 mg sodium, 35 g carbohydrates, 2 g fiber, 10 g sugar, 9 g protein, 4% vitamin A, 15% calcium, 70% vitamin C, 6% iron

lemon berry
SMOOTHIE

1 cup water

1 tablespoon fresh lemon juice

1 teaspoon sugar

1 cup strawberry frozen kefir (such as from Lifeway)

This is a light, refreshing probiotic-rich drink—try adding 1 cup of fresh sliced strawberries for an extra nutrient boost. SERVES 1 (1½ CUPS)

directions

Combine the water, lemon juice, and sugar in a glass or small mixing bowl. Stir to dissolve the sugar and pour into the blender. Add the strawberry frozen kefir and pulse to blend.

Serve, or store in the refrigerator for up to 3 or 4 days.

nutrition facts (per serving)

110 calories, 1 g fat, 0 g saturated fat, 0 g trans fat, <5 mg cholesterol, 60 mg sodium, 23 g carbohydrates, 0 g fiber, 20 g sugar, 4 g protein, 0% vitamin A, 15% calcium, 8% vitamin C, 0% iron

blueberry
SHAKE

1 cup frozen blueberries

1 cup low-fat or nonfat milk

½ cup vanilla frozen yogurt with live and active cultures (such as Stonyfield Gotta Have Vanilla)

This super blueberry shake has only 150 calories and 3 grams of fat, with 6 grams of protein per serving. Plus it freezes well: Consider making a double batch and freeze extras in ice pop molds for a frozen treat. **SERVES 2 (1 CUP EACH)**

directions

Combine all of the ingredients in a blender and blend until smooth. Divide between 2 glasses and serve immediately.

nutrition facts (per serving)
150 calories, 3 g fat, 1½ g saturated fat, 0 g trans fat, 0 mg cholesterol, 95 mg sodium, 26 mg carbohydrates, 2 g fiber, 17 g sugar, 6 g protein, 6% vitamin A, 15% vitamin C, 25% calcium, 2% iron

blueberry
CHOCOLATE
SHAKE

1½ cups light almond milk or Homemade Almond Milk (page 49)

1 cup chocolate frozen yogurt (such as Stonyfield After Dark Chocolate)

1 cup frozen blueberries

Blueberries and chocolate go great together and create a delicious icy treat. Each serving has just 120 calories. SERVES 2 (1 CUP EACH)

directions

Combine all of the ingredients in a blender and blend until smooth.

Divide between 2 glasses and serve, or store in the refrigerator for up to 3 or 4 days.

nutrition facts (per serving)
120 calories, 1½ g fat, 0 g saturated fat, 0 g trans fat, <5 mg cholesterol, 150 mg sodium, 24 g carbohydrates, 2 g fiber, 20 g sugar, 3 g protein, 8% vitamin A, 40% calcium, 10% vitamin C, 2% iron

creamy
RASPBERRY
SHAKE

1 cup frozen raspberries

1 cup low-fat or nonfat milk

½ cup vanilla frozen yogurt with live and active cultures (such as Stonyfield Gotta Have Vanilla)

This shake is a spot-on blend of frozen fruit, frozen yogurt, and milk. To work in the probiotics, choose frozen yogurt that has the words "live active cultures" on the package, like Stonyfield's frozen yogurt. In fact, Stonyfield even lists the specific live active cultures in their frozen yogurt: *S. thermophilus, L. bulgaricus, L. acidophilus, Bifidus*, and *L. casei*. SERVES 1 (1½ CUPS)

directions

Combine all of the ingredients in a blender and blend until smooth. Serve immediately.

nutrition facts (per serving)
270 calories, 3 g fat, 1½ g saturated fat, 0 g trans fat, 15 mg cholesterol, 170 mg sodium, 47 g carbohydrates, 8 g fiber, 37 g sugar, 14 g protein, 10% vitamin A, 45% calcium, 50% vitamin C, 6% iron

mint cookies-
AND-CREAM
SHAKE

1 avocado, pitted and peeled

½ cup coconut milk beverage (such as from So Delicious)

½ cup unsweetened cultured coconut milk (such as from So Delicious)

1 tablespoon agave nectar

1 tablespoon unsweetened cocoa powder

4 or 5 ice cubes

4 coarsely chopped mint cream cookies (such as Newman-O's Hint-o-Mint)

The majority (about 15 grams) of fat in this shake comes from avocado and it's healthy monounsaturated fat. Unlike most traditional shakes, each serving also packs in 12 grams of fiber (again, mostly from the avocado). It is certainly a dessert, but it's a healthier way to enjoy an icy treat! SERVES 2 (¾ CUP EACH)

directions

Combine the avocado, coconut milk, cultured coconut milk, agave nectar, and cocoa powder in a blender and blend until smooth. Add the ice cubes and blend until icy and smooth.

Divide between 2 glasses, stir half of the chopped cookies into each, and serve.

nutrition facts (per serving)
390 calories, 24 g fat, 8 g saturated fat, 0 g trans fat, 0 mg cholesterol, 95 mg sodium, 44 g carbohydrates, 12 g fiber, 31 g sugar, 5 g protein, 6% vitamin A, 15% calcium, 45% vitamin C, 6% iron

kombucha root
BEER FLOAT

½ cup Original Kombucha (page 23) or GT's Enlightened Organic Raw Original Kombucha

¼ teaspoon root beer extract

½ cup vanilla frozen yogurt (such as Stonyfield Gotta Have Vanilla)

This is an absolutely delicious twist on kombucha: The strong flavor of root beer extract with the bubbly texture of kombucha is perfect! Stirring in vanilla frozen yogurt (and its probiotics) adds the familiar creaminess of a root beer float. The only marked difference is that the liquid has a lighter color because kombucha is naturally much lighter in color than traditional root beer. **SERVES 1 (ABOUT 1 CUP)**

directions

Stir together the kombucha and root beer extract in a glass. Add the vanilla frozen yogurt and serve.

nutrition facts (per serving)
120 calories, 0 g fat, 0 g saturated fat, 0 g trans fat, <5 mg cholesterol, 70 mg sodium, 24 g carbohydrates, 0 g fiber, 20 g sugar, 4 g protein, 0% vitamin A, 15% calcium, 0% vitamin C, 0% iron

Resources

GADGETS

Greek Yogurt Maker
Dash, bydash.com

Yogurt Maker (aka incubator)
Cuisinart, cuisinart.com

Kombucha Brewing Supplies (brewing kits, bottles, and SCOBY)
Cultures for Health, culturesforhealth.com
Kombucha Brooklyn, kombuchabrooklyn.com
Williams-Sonoma, williams-sonoma.com

High-Speed Blenders
Ninja, ninjakitchen.com
Vitamix, vitamix.com

INGREDIENTS

Almond Milk
Dream, tastethedream.com
Silk, silk.com

Carrot Juice
Bolthouse Farms, bolthouse.com

Chai Tea Concentrate
Oregon Chai, oregonchai.com

Chia Seeds
Bob's Red Mill, bobsredmill.com
Nuts.com

Coconut Milk and Cultured Coconut Milk Products
So Delicious Dairy Free, sodeliciousdairyfree.com

Coconut Palm Sugar and Coconut Water Powder
Big Tree Farms, bigtreefarms.com

Dill Pickles (naturally fermented)
Bubbies, bubbies.com

Filmjölk
Siggi's, siggisdairy.com

Flax Oil and Fish Oil
Barlean's, barleans.com

Ginger Beer
Reed's, reedsinc.com

Granola
KIND Healthy Grains, kindsnacks.com

Greek Yogurt
Cabot, cabotcheese.coop
Chobani, chobani.com
Powerful Yogurt, powerful.yt

Kefir
Lifeway Kefir, lifeway.net

Kefir, Coconut Water
Inner-Eco, inner-eco.com

Kefir, Water
Caveman Foods, eatcavemanfood.com

Kombucha
GT's, synergydrinks.com

Miso Paste
Eden Organics, edenfoods.com
Great Eastern Sun, great-eastern-sun.com

Protein Powder, Hemp
Navitas Naturals, navitasnaturals.com

Protein Powder, Pea Isolate
Naturade, naturade.com

Protein Powder, Whey
EAS Sports Nutrition, eas.com
Klean Athlete, kleanathlete.com

Probiotic Juice Drink
Good Belly, goodbelly.com

Soy Milk
Silk, silk.com

Spirulina
Nuts.com

Starter Culture Resources
Cultures for Health (kefir grains, yogurt starter, SCOBY),
    culturesforhealth.com
Euro Cuisine (yogurt starter), eurocuisine.net
Kombucha Brooklyn (SCOBY), kombuchabrooklyn.com
Yogourmet (yogurt starter, kefir starter), yogourmet.com

index

Page numbers in *italics* indicate illustrations.